The Politics of Bodies at Risk

The Politics of Bodies at Risk

The Human in the Body

Maria Boikova Struble

ROWMAN &
LITTLEFIELD
INTERNATIONAL
London • New York

Published by Rowman & Littlefield International, Ltd.
6 Tinworth Street, London SE11 5AL
www.rowmaninternational.com

Rowman & Littlefield International, Ltd. is an affiliate of
Rowman & Littlefield
4501 Forbes Boulevard, Suite 200, Lanham, Maryland 20706, USA
With additional offices in Boulder, New York, Toronto (Canada), and London (UK)
www.rowman.com

British Library Cataloguing in Publication Information
A catalogue record for this book is available from the British Library

ISBN: HB 978-1-78660-122-3

Library of Congress Cataloging-in-Publication Data Available

ISBN: 978-1-78660-122-3 (cloth : alk. paper)
ISBN: 978-1-78660-124-7 (electronic)

∞™ The paper used in this publication meets the minimum requirements of American National Standard for Information Sciences—Permanence of Paper for Printed Library Materials, ANSI/NISO Z39.48-1992.

Printed in the United States of America

Dedicated to the memory of my brother, Ivan

Contents

Acknowledgments

I am grateful for the opportunity to work on this project, which allowed me to fall in love with my discipline again. Thank you to Dhara Snowden, Natalie Linh Bolderston, Gurdeep Mattu, Hannah Fisher, and various members of the editorial, marketing, and graphics teams at Rowman & Littlefield International, who walked me through the process of completing this manuscript and who believed in me enough to trust that I could finish it, even when my sentences remained too long and when time for revision was short. I am grateful for financial support from the Professional Activity Fund at Western Colorado University, which made my travel to Bulgaria and Nepal possible. I wish to recognize the library staff at Western Colorado University for help with research and for locating long lost dissertations in Indian University libraries. Thank you to my colleagues at the Department of Behavioral and Social Sciences, without whose support I would certainly have lost my mind teaching four courses while writing this book. Thank you to Donna, Heather, and Suzy for being friends on top of being colleagues and for helping me get through this simply by not letting go. I am especially thankful to my colleague in the Politics and Government program, Dr. Brian Bernhardt, whose endless cheer and good will, intellectual support, hot meals, open-door conversations, and holding down the fort while I was on sabbatical leave made this possible. Brian, I am honored to be working with you.

Thank you to my students at Western Colorado University throughout the years for entertaining my "thinking outside the box" and who, especially in the spring 2018 semester, waited patiently as I canceled office hours so I could write, cheered me on when I made progress, and gave me hope when I was struggling. I know you all said you wanted to read this, but I assure you, the process of writing remains more exciting than the final product. Thank you especially to Taylor Cull, Keriann Conroy, and Stefano Ballesteros, who make teaching the best job in the world.

Thank you to everyone in Bulgaria who collaborated and provided their time and energy as I was struggling with the bureaucracy of my home country. Antoaneta Vassileva was singularly inspiring and a good friend. Thank you for reminding me that even if it's only one life we are helping, it matters. The staff at P.U.L.S. in Pernik opened their doors to me several times and embraced my ideas and also provided useful critique. Cvety Naidenova of IOM-Bulgaria was by far my biggest cheerleader—Cvety, thank you for the dates, coffee on the fly, and for opening all the doors for me, literally and figuratively. On many an occasion your sheer willpower and belief in me were the only things that kept me going. I am in your debt. I am grateful to all survivors of trafficking in human beings who agreed to speak with me, sometimes under challenging circumstances, and whose generosity, wisdom, and singular experiences informed not only the direction of this project but also the direction of my research as a whole.

In Nepal, I cannot thank Nabyn and Reena enough for many hot dinners, for taking care of me in their home, for their love, care, and friendship, and for making me laugh. The many Tibetan refugees who shared their experiences and bits of history continue to live on in the pages of this book. Their resilience, compassion for others, belief in "happy endings," and kindness are unmatched, and I am lucky to have had the opportunity to hear their stories. The children at Atisha Primary School in Jawalakhel Tibetan Refugee Settlement in Kathmandu brought so much joy. *Thuk Je Che*! Thank you to the teachers at Atisha Primary School, who took me under their wing, fed me, went with me on walks in the rain, helped with interpretation, and became my friends. Sonam, Tashi, Kalsang, Tsering, Sumjok, this is for you.

My academic mentors remain Dr. Michael Dillon and Dr. Roxanne Doty. Here is to finding my bearings, Mick; here is to being present, Roxanne.

Many friends stood by me in various ways as I was working on this project: Amanda Birdsong, Nicol Talbert, Molly Mugglestone, Laurie Boscaro, and Christina Buchanan in Gunnison, Colorado; Tsvetanka Radovanova, Silvia Prokopieva, and Elka Babacheva in Bulgaria. Thank you for watching the boys while I was writing, for making me go out for drinks, and for random acts of kindness. I promise to answer the phone more often from now on.

I would like to acknowledge my family: thank you to Candace and Edwin Higginbotham for their love and undying belief in the ability of their adopted daughter to do great things, even when she did not believe that herself; Coreen Spencer and Gary Heisler for their love and for teaching that peace matters; Jim and Betty Struble for cheering me on all the way from the Upper Peninsula of Michigan and for allowing me to confide in them when I needed to; Desi, who always had a warm bed for me in Sofia and whose heart spans continents; and my parents, Vanya and Boyko, who took care of the boys in Bulgaria while I left on research trips for weeks at a time, who did not always understand what I was doing but never questioned that it mattered to me. I thank them both for worrying whether my clothes were ironed, whether the car had enough gas, whether I had finished my lunch, and for never stopping me from taking the journey.

My children, Damien and Miro, can be found on every page. I thought of the kind of world I wanted them to inherit as I wrote this. I am humbled by their love, patience, and reminders that taking time to play Legos is one of the most rewarding things a parent can do. Thank you for cheering me on even when you missed me as I was gone at bedtimes and mealtimes working on this project. You are beautiful, wise, and my world; I am the luckiest mama in the world.

Darren, my life's partner, has been the single greatest inspiration behind this book. Even when it seemed like it wasn't going to get finished, I could not fail him—so I finished it. For his unfailing belief in my ability to tell a different story, for his singular love, critical insight, long runs on cold December mornings, and holding my hand no matter what part of the globe I was in, I am forever grateful.

My brother, Ivan, did not live to see this finished. I know he would have been proud of me; he was always my biggest fan. I dedicate this book to his memory.

Chapter One

Risk and the Exception

Sovereign Power and Its Excesses

This project begins with the proposition that the nature of sovereign power today can best be understood as a preoccupation with managing risk and insecurity. What used to stand for the ultimate expression of sovereignty, namely the "production of general norms by a body (the demos) made up of free and equal men and women,"[1] has been replaced instead by various means and measures of governing insecurity, in which the referent object of security remains the human body but also the idea of security itself. Specific idealizations and operationalizations of risk are discussed in the context of experiences of individuals who have lived through insecurity, which is understood most especially in the context of crossing an international border. In this sense, not only will traditional conceptions of state security be interrogated and shown as unstable and evolving constructions but the idea of the "border" as a territorial demarcation will also be examined by looking at measures taken by sovereign nation-states to perform, iterate, and solidify representations of security in a climate of international, national, corporeal, military, economic, social, and environmental threats.

At its core, international relations (IR), when expressed as the study of the interactions between sovereign nation-states, occupies itself with questions of power and its dissemination. Importantly, the relations between nation-states have also necessitated certain rules of engagement if only so that unnecessary conflict and potentially deadly wars could be avoided. In the midst of all this is the pervasive and equally unquestioned imperative to live up, in a truly Hobbesian sense, to the obligations of the "social contract" through measures aimed at ensuring the survival of the population. The latter is defined as those on the inside, citizens and legal residents, against the many variations of their illegal counterparts, enlisting in the process the cooperation of the former in efforts to mitigate the anticipated negative consequences from not being entirely in control of the latter. Those on the inside, as a result, are encouraged to support legal or extra-legal measures of surveillance. What becomes of our understanding, faith in, and idealization of sovereignty when the latter affirms itself through the general "*instrumentalization of human existence and the material destruction of human bodies and populations*"?[2]

Taking lead from the work of Michel Foucault on governmentality, which makes "visible, through a particular perspective in the history of the present . . . the different ways in which an activity of art called *government* has been made thinkable and practicable,"[3] the project endorses Foucault's views on power as the cumulative effects of relationships of oppression and resistance to that oppression.[4] This is opposed to a traditional view of power as mere domination/authority. This project draws conclusions regarding the exercise, manifestation, and performance of said power with regard to native and nonnative populations of migrating humans. The concept of risk is useful for a discussion of sovereign power and its excesses as it helps situate traditional sovereign power deciding on life and death vis-à-vis governmental power in which "the law becomes subordinated to the logic of government, deployed as a tactic in the administration of things in order to optimize the well-being of a population."[5]

The discursive nature of Foucault's work invites readers to engage with the historically specific ways in which power works and is disseminated. There are many unexamined examples of everyday power struggles that not only help shed light on discussions of the state and security within IR but also help think through how the body of the migrant as the latter's authentic singular story navigates power and is navigated by the "endless and open strategic game"[6] that are its practices. Said practices of "everyday security" are found in the intersection between making life live and trauma, specifically with reference to the lives of survivors of trafficking in human beings and refugees. The precariousness of said populations forces them to position themselves vis-à-vis sovereign apparatuses, governmental and nongovernmental organizations. The project examines what the place, relevance, and importance of "everyday security" is alongside sovereign practices of management, government, and defense of populations.

There is an obsession within IR, mirrored by an obsession within the sovereign center, with the need to self-validate. It is operationalized as various wars on (in)security, vulnerability, (in)stability, and (dis)order and has almost theorized away indeterminacy, threat, and danger altogether.[7] This has, as this project will argue, ironically only increased nation-state and individual insecurities and instituted a reality in which exceptions, emergencies, and ruling outside the law have become the norms according to which the sovereign is understood. Whether we are talking about commuting by train, crisscrossing the world by airplane, living in sea-level urban metropolises, going to a concert, climbing Mt. Everest, surfing the internet, shopping for food, buying a car, attending school or a wedding, or going to the market, we are constantly reminded of the myriad dangers lurking from within and without. The omnipresence of said dangers has produced a parallel universe of measures designed to tackle them, a universe invested above all in its perpetual renewal and self-preservation, in the reiteration of its pertinence, relevance, and invaluability. Loss of jobs, retirement savings, medical coverage, affordable housing, means of gainful employment, and opportunities for upward mobility all add to the list of "natural" disasters, yet another list—one of dangers and insecurities incurred in modernity and perpetuated by our own success at mastering the (natural) world, pushing human, technological, and economic progress, invention, and ingenuity ever further and further forward while neoliberal success births a bagful of dire and "risky" consequences.[8]

The new terrorist-induced fears and concerns over securing borders, bodies, nations, and "our way of life" have validated the modern nation-state, as terrorist and other territorial threats proliferate and continue to shake up the territory-citizen-border foundation of the nation.[9] What the state has done in response has been to buttress its efforts at constructing and disseminating a discourse explaining (in)security to its citizens and then proceeding to do security-like things to manage insecurity. States perform acts of security for their citizens and others watching through various legal, extra-legal, and even illegal measures aimed at erasing or at least arresting the visibility of threats, the permeability of borders, and the sterility of security promises.[10] Risks are not necessarily averted, nor are they preemptively addressed in any meaningful way, but their proliferation is tolerated and even celebrated in the presence of measures and procedures aimed at addressing the fallout of such risks. These include state-sponsored rehabilitation programs, monetary support, military and border enhancement, economic and other treaties, or policies informed and underwritten by a larger universe of supra-national and other international agreements, treaties, and partnerships. Rather than recognize that living with insecurity is part of modernity, states work tirelessly to erase the impression of said insecurity, and in the process erase experiences, trauma, and even lives.

Risk-informed insecurity, in this sense, redefines the nation-state project of security as an instance of governmentality understood as the art and practice of governing through mechanisms of security away from a social contract aimed at bringing people together. This art validates practices of state power meant to delineate, separate, and, if necessary, execute. Thus comes the inauguration of a legitimate institutional reduction of the "multiplicity, complexity and conflictuality of identifications and senses of belonging" through the application of a "preventive violence or a 'symbolic' or material—corporeal—organized counter-violence"[11] in a process of "annihilation of difference"[12] that often works to reaffirm itself. Liberalism, in this sense, can be understood not so much as a doctrine or political ideology but "as a style of thinking quintessentially concerned with the art of governing."[13] In the context of this project, said art of governing is interrogated through an examination of practices governing insecurity, starting from the Foucaultian thesis that "the finitude of the state's power to act is an immediate consequence of the limitation of its power to know."[14] This project understands performances of security as modalities of knowing, examined here in the contexts of human trafficking, the experiences of Tibetan refugees, and children's deaths in war.

It is by examining practices that refuse to accept and subject the individual to such annihilation(s) that one can highlight the productive relationship of sovereign power vis-à-vis the individual human being. Rather than debate power through a focus on the poor, the dispossessed, the illegals, and the criminals among us, which are its usual subjects, one can begin to interrogate the ways in which power works on us all. This is what the experiences of survivors of trafficking in human beings and refugees, to name two examples engaged with in this book, uncover and why they are relevant beyond their immediate circumstances. In this sense, "people displaced by capital with the aid of states that claim to protect their rights exist on a continuum with stateless people" and can be read as inhabiting "gray zones in which life becomes matter—where no form of law seems to offer any protection, where a blind eye is actually turned, and where the disposability of peoples becomes not exceptional, but the byproduct of capitalist excess."[15] Sovereignty in the register of making live or letting die is recast as "emergent governmentality" where the factors that underwrite the sovereign's ability to care for or dispose of human life are understood in the context of "disposability" and legitimatized by the ability of said sovereign to "withdraw any means through which this life and death could be administered."[16]

I examine and showcase some of these ameliorative, divisive, and excessive practices tested on so-called disposable people and on legal, legitimate subjects themselves, with specific attention to individuals and groups of individuals defined by their passive or active relationship to borders and their regimes. The border, in the context of this study, is understood as the real, material, and corporeal territorial demarcations of nation-states meant to delineate inside from outside, domestic from foreign, citizen from alien. But borders are also discussed as a condition of flux, as evolving and idealized spaces that would not be "conceived of as the support of the universal, if they were not imagined as the point at which 'world-views,' and thus also views of man, were at stake: the point at which one must choose, and choose *oneself*."[17]

If, as Étienne Balibar maintains, "the nation is, ultimately, an *impossible* entity, which can never entirely achieve its ideal, and it is as such—that is to say, as a problem—that it is *real*,"[18] how does it keep up the ruse of legality while proceeding to punctuate identities, define belonging, and exercise so-called legitimate police, juridical, and military power unabated? Through the mutual interplay between discourses of (in)security and risk, wars (on drugs, poverty, terrorism, trafficking, migration) are no longer waged in order to protect from a scary, unknown enemy but on behalf of demarcated and highly segregated groups of individuals belonging together by virtue of being similar. The sovereign power to wage war is thus supplemented by the ever-emerging power of the nation-state to regulate life in validating "the total mobilization of social forces for a united purpose that is typical of warfare,"[19] in which the liberal promise of rights and equality is subsumed under the rubric of state practices of security. This liberal promise is thus divorced from the checks, balances, and equality that voters are led to believe in. This "emergent governmentality" of which I speak is characterized by the "administration of the absence of order"[20] allowing for the "rule over or management of the permanent state of exception."[21] The cases of trafficking survivors, refugees, and children dying while trying to cross an international border, rather than underwrite the workings of this "permanent state of exception," underscore its inability to address said singular, unique experiences in any meaningful way without relegating them to exceptionality. The ensuing chapters show how this happens and the practices of resistance that emerge in response.

OVERVIEW OF THE ARGUMENT

The book unfolds as follows. Chapter 1 introduces the concepts of risk and insecurity and positions them within IR and existing scholarship on the subject. While security remains the main measure of state success, the chapter proposes that we understand the state instead as another example, alongside refugees and trafficking survivors of uncertainty and fear. Outside political machinations and party lines, the fundamental nature of the nation-state as such always exists in the spaces between order and disorder and is born in the clash between civilization and barbarism, the self and the other, security and chaos. In fact, I propose that terror is not alleviated but born within the state, and the exercise here of interrupting said practices allows us to embrace trauma, remembering, and risk as full-fledged part of practices of governing.

Chapter 2 provides a theoretical overview of Giorgio Agamben's writings on the exception and "bare life" as they will serve as the main bearing point against which to position the case studies engaged in this book. In light of current migration, border, and insecurity crises around the world, Agamben's work is overdue for a novel engagement and reinterpretation against the "emergencies" facing us today. Concepts of risk and insecurity are recast against Agamben's well-known engagement with the norm and the exception as the referent objects of sovereign security. Agamben's work not only provides an avenue into thinking about how to understand instances of sovereign violence but also, precisely because of its omissions, inaugurates possible ways of protesting and addressing said violence.

Chapters 3 and 4 engage the experiences of survivors of trafficking in human beings. Chapter 3 provides an overview of the literature on human trafficking, with specific focus on how it has been understood and discussed in relation to questions of (in)security, interrogating existing definitions and statistics relating to trafficking in human beings. It positions chapter 4 within the existing discussion on the subject and provides a brief overview of the most recent numbers relating to trafficking worldwide. The Bulgarian context is introduced also from the point of view of existing national directives, agreements, and examples positioning them alongside the 1989 United Nations Convention on the Rights of the Child. Chapter 4 examines the testimonies of trafficking survivors in the Bulgarian context. It then proceeds to engage with the concept of risk in the context of human trafficking, offering the reader what I consider a novel take on trafficking. The management of survivors and their needs uncovers a systematic pathologization, turning trafficking almost into a "disease" that, given the right kind of approach and remedies, cannot but be cured. Additionally, the chapter problematizes conceptions of security and risk understood to be produced *outside* the nation-state and reads them as consequences of insecurity *inside* the state, the result of which has been the casting of sovereign weakness as sovereign strength, already alluded to in chapter 1. Rather than engage with risk and insecurity as givens, the chapter traces their enactment on the bodies and experiences of survivors of trafficking in human beings. Survivors' testimonies are presented in the form of poetry. The latter allows for a novel engagement with the trauma resulting from having been trafficked, and for remaining committed to the process of bearing witness as a singular event. The writings of Jenny Edkins and Cathy Caruth help position my discussion vis-à-vis other traumatic events and with respect to already existing literature on the politics of trauma. I argue that the poetic form, in addition to allowing for a direct telling of survivors' testimonies, honors a commitment to engaging the subjects of our research from the point of view of uncertainty and our own undecided subjectivities, honoring in turn a process of oscillating among realities of strength, vulnerability, victimhood, and overcoming.

Chapter 5 focuses on the experiences of Tibetan refugees in Nepal. It continues the project's interrogation of insecurity incurred while crossing a border, but this time by looking at an example of protracted displacement through the metaphor of waiting. First, it overviews the history of Tibetan displacement briefly. The metaphor of "home" is discussed as it relates to yet exceeds our understandings of security, risk, and the nation-state, in this case with reference to the Nepali government, the Chinese government, and the Tibetan Government in Exile (TGiE). In a sense and much like trafficking survivors, Tibetan refugees are overgoverned, positioned between the strategic interests of several nation-states and feeling very strong commitment to the policies, philosophy, and teachings of His Holiness the Dalai Lama. Aware that the TGiE does not possess official political status worldwide, I make the case that it nevertheless acts like it does. Despite lacking recognized claim to political legitimacy and a firm control over demarcated territory, the TGiE has state-like status.

Additionally, what makes the Tibetan population different than most other refugee populations is the fact of their prolonged displacement, now in its sixtieth year. In this sense, the physical displacement so characteristic of refugee populations is here augmented and complicated by a temporal one, in which being in exile also becomes symptomatic of being "out of time" as the reality of displacement is more than a temporary inconvenience but has turned into the norm. In their daily dealings with the Nepali state, despite their lack of documentation and unrecognized in their refugee status, refugees engage in practices of sense making, bearing witness to examples of "everyday security" and the myriad ways of circumscribing the Nepali state in order to enact and realize belonging. The chapter moves back and forth between the realities and experiences of the Tibetan refugees in Nepal and the challenges of conducting research there.

Chapter 6 concludes my discussion of risk and insecurity inscribed on the bodies of individuals crossing international borders by engaging with experiences and the ultimate death of children victims of ongoing armed conflicts. It begins by discussing representations of the body at/in war, opening with World War I and ending with the War on Terror. Common perceptions of the "soldier body" are examined for their emphasis on heroism, courage, and valor. Parallel engagement with the wounded body at war problematizes the insistence on wholeness for which state narratives of the soldier body stand. The chapter insists on referring to children's bodies in war as "war bodies" and not "dead bodies." This change in naming is not insignificant and serves as the culmination of a call for a different way of addressing survivors of trauma that has informed the book so far. Children and trafficking survivors both exceed the label of "war bodies" and stand for more than victimhood, in a similar way that Tibetan refugees are more than just stateless people. The chapter offers a rereading of the 1989 United Nations Convention on the Rights of the Child, its history, content, and critiques, and culminates with a close reading of the experience of seeing and explaining the deaths of two toddlers, Aylan Kurdi and Fareed Shawky, both victims of the international War on Terror, as well as the war experiences of a third child, the Syrian boy

Omran Daqneesh. The discussion uses children's deaths in war as a frame for interrogating the implications war has for sovereign security and the ways in which bodies are understood, treated, and remembered by the state and by us all. The insecurity and vulnerability inherent in sovereign practices, as discussed in prior chapters, has existed as a sovereign claim over life. The children, in death, give us an opportunity to see security also as a sovereign claim over death. War as lived experience is engaged through the individual stories of the speechless, "ungrievable" reference objects of its violence, if only so we can lay them to rest and allow them and ourselves some dignity in the process.

The conclusion asks: Now that we have interrogated the state, engaged the inherent insecurities at its core, and thrust them aside so other voices, other stories, and other instances of "securing" can be embraced, should we secure the state another way? If the critique offered here is an example of how to "think otherwise," is it also a prescription for "being otherwise"? As every critique in itself gives birth to a resistance, what might the potential unanticipated effects of this move to unsettle the sovereign be? How might this move be perceived by the very subjects of this study, whose status challenges but also evokes a strong desire for eradicating this precariousness once and for all? Finally, what does the road forward look like for research interested in the in-betweens, lacunas, and silences that accompany insecurity? One path would be a continued insistence on uprooting the state. Another has us locked within the sovereign embrace. Giving birth to the human from there remains the challenge.

SECURITY AS PATHOLOGY

The argument here holds that fear, risk, and danger have become the prevalent modes of governing today. Regimes of governmentality by way of technologies, agencies, programs, and initiatives responsible for rehabilitating or "legitimizing" human beings' claim to inclusion work to normalize a kind of "state of exception" that is predicated on the perpetuation of insecurity and not one informed by a concerted effort at its eradication. What is more, and as a result of this reality, human life as such has become defined as the thing whose ongoing commitment to securing validates various "mechanisms . . . of fear management."[22] Said management has taken the form of a performance, in which the ongoing construction of security discourses for the benefit of the general public and those in charge has substituted, and continues to displace, sovereign measures aimed at doing what it takes to keep the people safe. It is this performative dimension of security that will be engaged in greater detail in the chapters to follow, especially with respect to the construction of security-scapes inside as well as beyond the borders of the nation-state. Last but not least, the ensuing pathologization and criminalization of feared and pervasive states of insecurity not only underwrite nation-states' claims regarding the legitimate control of modes of rehabilitation and order, but also the very idea of what an authentic presence means and who can be authentically present within the boders of a state. Security practices and discourses, I argue, erase authenticity, and the consequences of this erasure affect us all for the worse.

Neither trafficking victims nor Tibetan refugees in Nepal nor children drowned in the Mediterranean actually threaten the nation-state in any way that we are used to thinking about insecurity. They do not represent examples either of a terrorist threat or of hostile enemy presences, yet they are discussed as if they do. While they are not direct threats to the physical security of the nation-state, they uncover a sovereign uncertainty and inability to narrate and provide security. This bears real implications for people's ability to lead normal, dignified, and secure lives, in some cases even precipitating their deaths. The migrant, the trafficking survivor, the Tibetan refugee, and the dead or drowned child exemplify states of insecurity that are inherent in human existence, yet understood as anomalies, justifying the creation of a "security state." As a result, the perpetuation of insecurity somewhere perpetuates insecurity everywhere, highlighting "the modern state as essentially *managerial*."[23] The latter recasts its failures to provide security into tasks to be managed and, by extension, manufactured, much like the "manufactured uncertainties [that] make society more reliant than ever on security and control."[24]

I argue that the reaction of states with respect to anticipated threats is always done as a knee-jerk reaction: for example, when someone explodes themselves with liquid explosives, there ensue prohibitions on carrying liquids on the underground, on trains, and on other modes of transportation. But what happens when Tibetan monks self-immolate? What happens when the Chinese authorities put out the fire and disappear said monks? When a terrorist attack occurs on U.S. soil, stricter travel restrictions are immediately imposed, with the bulk of the consequences affecting the movement of people who are different and not immediately part of the American Dream. In other words, what these examples show is that global risks remain hopelessly localized in their nature with terrorism not registering as such on U.S. soil until it becomes "terrorism on U.S. soil." Lacking adequate means of addressing such pervasive global dangers/risks, the nation-state proceeds to engage in "knowledge politics" in which the "individual is subjected to power, by being circumscribed ('measured') as an object of knowledge . . . [in which] the body is controlled by being *known*."[25] Put another way by Claudia Aradau and Rens van Munster, what is considered risk are things that "*become* risks through the way that they are rendered knowledgeable and thinkable as well as through the technologies that are mobilized to tackle and manage them."[26] This way of constructing security bears on a performance where "risk inscribes reality"[27] so much so that "the identification and management of risk is a way of organizing reality, disciplining the future, taming choice and rationalizing individual conduct."[28]

Taking a step back from prevalent discussions on the role of biometric security in executing the promise of nation-state security, the chapters in this book engage the migrating body as a referent object of power. The latter exists in the spaces between "security-as-sovereignty," which largely deals with borders, territory, and the state itself, and "security-as-governmentality," which largely deals with bodies of people and life through technologies of biopower. As a result, the modern subject of risk society lives between the two, constantly traversing "security-as-sovereignty" and "security-as-governmentality" and thus proliferating "self-monitoring and self-carcerating techniques that can only confirm that they [populace] have good reasons to remain fearful."[29] Ironically, sovereign power as "hierarchy" is largely unable to address the challenges, as outlined above, that face modern existence.

It is no longer enough to apply a "law to an object," nor does security neatly "flow down and out from a central sovereign source" where "its primary form is prohibition."[30] A strategic account of power "is understood instead in terms of formations of relation that are amenable to manoeuvres that have specifiable productive effects in terms especially of objectification and subjectification."[31] The result is a situation where Ulrich Beck's "relations of definition" explain "the concrete changing soul in which the sovereign power is grounded, the conditions which make it possible for it to function"[32] and "reproduce a life that is amendable to its sway" so that it remains the "strategic ordering of life."[33] Here, life becomes a malleable and changing means to power's own continuous reproduction, in this case through the play of "risk discourses." Coming full swing to Wendy Brown's discussion of the loss and bastardization of freedom as a meaningful liberal concept,[34] resistance as conceptualized by Foucault becomes the state of moving away from, around, and through regimes of power in a meaningful iteration both of authenticity and of freedom.

Said otherwise, if the modern task of government has become to make people uncertain and anxious, what is people's response to this effort to redefine, combat, and even rethink what this uncertainty and anxiety imply? What does it mean "to speak in one's own name not as a manufactured individual made up of a variety of purchased or pre-packaged identities?"[35] Evoking a politics of relation as a politics of authenticity, in the context of this study, opens the door for speaking "in one's own name" so that "enabling relationships sustain us in the ambivalence of our responses"[36] away from what Louise Amoore calls "proxy forms of sovereignty" as expert knowledge, consulting, and the like.[37]

If, after Roxanne Doty, there is no such thing as the state but only a powerful desire for the state that "underlies the various productions of the inside realms of social protection, governmentality and order as distinguished from the outside realm of anarchy and disorder,"[38] how are specific practices of statecraft, as the seminal and quintessential manifestations of state power, informative of the state's obsession with security or rather with eliminating insecurity as the antithesis of legitimate sovereign order? How are specific practices and tools of governmentality with respect to dealing with and addressing the needs and realities of survivors of trafficking in human beings, migrants, and children at war instrumental in the legitimation, performance, idealization, and construction of sovereign power? Here, the register of the exception, especially as discussed by Carl Schmitt,[39] can best be understood as being validated by the occurrence of an unusual, extraordinary, singular event that exceeds the register of existing law. In this sense, the exceptional as an originary expression of sovereign power is the byproduct of recognizing and honoring a type of sovereign singularity that Daniel McLaughlin calls "total war," Michel Foucault calls "biopolitics," and Giorgio Agamben calls "the state of exception." In the context of learning what wounded, victimized, and violated human beings can teach us about what politics is, Françoise Vergès rings true that moralistic condemnation of violence is a weak way of avoiding future violence, especially as we find ourselves caught in a framework of "identification with victims" producing "righteous, pure and innocent identity."[40] This project's engagement with personal stories reminds us, without falling prey to fear, that the disorder of everyday life is inherent in the human condition, and that said disorder, when used to validate fantastical state interventions, becomes dangerous.[41]

In the same way in which "the state of exception" has served to constitute risk and insecurity as a master unified discourse, so are human experiences and histories also subject to a process of constant reductionism underwriting identity formation, as if there were such a thing as the core self, as if there were such a thing as the state's core essence. But if "there is no identity which is 'self-identical' and since *all identity is fundamentally ambiguous*,"[42] the political apparatuses and mechanisms responsible for securing said "unified identities" also come into question. It is important to recognize that any attempt to define power away from its multiple effects obscures the fact that each apparatus of power yields a different register of control over individuals/groups, producing "different effects . . . different kinds of possible resistance, and . . . a different analytical frame."[43] Specific reference to the meaning, workings, and securitizing practices with regard to the discourse on citizenship as the ultimate "legitimizing political act"[44] sheds light on the ambiguous nature of borders as well.

By the ambiguous nature of borders I do not mean that borders are not a real thing. Built fences and security checks at international crossing points bring about real consequences for travelers. Benjamin J. Muller shows the real-life effects that regimes of border security had for one Polish citizen, Robert Dziekanski, who died at Vancouver International Airport as a direct result of "the flagrant overuse of the Taser" but also as an indirect result of a catastrophic failure by the Canadian state to anticipate an out-of-the-norm situation and, in the process, provide security to travelers.[45] Giving testimony to lived experiences must be understood in the context of said lived experiences inhabiting a reality no "larger and longer than the moments of the lives they speak from; they cannot be anointed as 'authentic' or 'true' since the experience they announce is linguistically contained, socially constructed, discursively mediated, and never just individually 'had.'"[46] To that end, the testimony of the subjects engaged with in this project inhabits a universe allowing for multiplicities of meanings existing alongside the realization that "truth has been established as the secret to our souls not by us but by those who would discipline us through that truth."[47]

The project takes its lead from Étienne Balibar's pronouncement that "what can be *demarcated, defined* and *determined* maintains a constitutive relation with what can be *thought*,"[48] and as such unthought. The disposable lives and human waste they incur informs the focus of the work presented here. Said otherwise, "to dis-pose is an exertion of power by the disposer in the utterance of a disposition made explicit, or, indeed, in a decision to exercise control over or a discarding of bodies."[49] Rather than thinking of nation-state security as defined by demarcating borders in the context of the multiplicity of modern transnational movements of people, goods, and capital, the border as a territorial, material entity subject to "ongoing contestation and negotiation"[50] is reenvisioned, rejected, or, in the least, rethought.

Risk understood to a large degree as unpredictable, unknowable, and hard to anticipate and mitigate is nonetheless explained in the context of tangible and specific state actions that are, importantly, reduced more and more to practices of management and programs "for the elimination of violence."[51] This requires the obedience of the subject and the simultaneous proliferation of multiple regimes of control directed at reducing complexity, indeterminacy, and risk through the exercise of legitimate violence, even cruelty.[52] "Thus, the effort to 'outlaw' social injury powerfully legitimizes law and the state as appropriate protectors against injury and casts injured individuals as needing such protection by such protectors."[53] The result is a situation in which the use of violence is legitimated as a measure and a tool for countering violence/risk/insecurity, "which have destroyed or disturbed an originally ideal, originally peaceful, non-violent order,"[54] or at least the idea of one. This anticipates, in effect, the proliferation of violence and, ironically, encourages its perpetuation, "making the perpetrator hurt as the sufferer does"[55] and reducing politics to punishment.

The seeming irony here nonetheless helps explain the legitimating role the subject plays in its obedience as well as resistance to nation-state practices and discourses of control. The goal of what follows is to examine this tension in greater detail and with reference to the specific cases of individuals and groups of individuals caught once in a network of risk-induced behavior and then in a world of risk-informed rehabilitation, implying them in a logic designed to reestablish order and ensure survival, in which members of the legal/legitimate national community are dis-membered, remembered, and unmembered either because of ethnicity, poverty, borders, warfare, or a combination of the four, yet rarely because of their histories, names, or singular lives. The project takes its lead from Wendy Brown's insight that as "subjects and practices are always at risk of being resubordinated through the discourses naming and politicizing them,"[56] as said discourses function within mechanics of governmental and disciplinary power,[57] so does their theorization always contain the possibility for resistance and, in resisting, for freedom. The cases examined engage the question of sovereignty as the exercise of "control over mortality"[58] in which life and living in the spaces between the exceptional registers of death inform authenticity. These registers are understood as the forceful, powerful, and sometimes illegal execution of a sentence on the part of a sovereign center vis-à-vis a domestic criminal, an enemy combatant, or a foreign subject who has knowingly or not transgressed the register of acceptable behavior at the border.[59]

The specific questions the project engages are: How are individuals who have lived through insecurities and risks as a result of crossing a border handled and reconstructed by the nation-state? How do they understand their place in an economy of risk making, risk taking, and risk unmaking that implies them in its workings once as legal, another time as illegal subjects? In what ways do their actions shed light on academic and legal understandings and commentaries on "the state of exception" as an important category of belonging and delimitation? How do said individuals problematize their own relationship with a so-called rehabilitative state apparatus? In other words, can the subject, an assumed carrier of rights, be it a citizen, an illegal, or a minor, think security without being constantly at risk of falling prey to an economy of exclusion, oppression, and fear? What, by extension, is the fate of human life in an environment continuously defined and usurped by the powers of "world risk society" and insecurity concerns of varying sorts? Could the stories of individuals having to negotiate risks and insecurity, both while crossing borders and inside them, help inform our understanding of the ways in which we, so-called secure citizens/subjects, position ourselves and our own security, how we move within a universe of widening class divisions, terrorist threats, and eroding human plurality? Last but not least, how are the deaths of children as a result of war the decisive moment of modern politics, echoing Agamben's pronouncement that the camp "is the new biopolitical *nomos* of the planet"[60] in which the production of bare life legitimizes sovereign power and its ability to produce and normalize a "state of exception?"[61]

I take lead from Ulrich Beck's insight relating to the uncontrollability of risks resulting from our own success[62] and bear in mind the ever-present terrorist threat we are exposed to by virtue of post-9/11 security logics. Examples of trafficking in human beings, protracted refugee situations, and children dying while attempting to cross the Mediterranean sea in makeshift boats inform parallel engagements with risk and insecurity from scholars and others not directly involved in combating terrorism or the next uncontrollable flood. Prevailing constructions and definitions of risk inform the ways in which we think, address, and conceptualize (in)security lived by ordinary people. Yet the apparatuses of the nation-state otherwise responsible for providing security, enabling citizens to live productive lives, and ensuring their protection also undermine rights the so-called social contract is supposed to bestow on citizen subjects.

On the one hand, Ulrich Beck, Anthony Giddens, and Wendy Brown argue that because of the inability of nation-states to address the myriad current risks the globe faces (environmental, terrorist, economic, and man-made natural calamities, to name a few), the imperative becomes to devise and commit to a cosmopolitan orientation in tackling them.[63] "World risk society compels the nation-state to admit that it cannot fulfil its self-declared constitutional promises, namely, to guarantee its citizens what is arguably the highest legal good, namely, their security."[64] On the other hand, we are enmeshed in a reality defined and shaped by terrorist and other threats our governments and all other entities of sovereign importance are feverishly, if unsuccessfully, working to eradicate. Following their persistent failure comes an insistence on having to employ novel tactics, deploy more boots on the ground, reenvision the use of smart technologies, and so on. This is the performative, spectacle-like dimension of security.

In this sense, it is useful to remind ourselves that so much of what the sovereign does to validate its hold over the means of power, control, and discipline are bureaucratic and "rehabilitative" functions, though sometimes also prohibitive ones in the example of wall building, aimed at restoring a perception of security in the populace. Trafficking is one example of such bureaucratic functioning, and so is burying the bodies of soldiers killed in combat. The bureaucratic commitments of states can only deliver on a limited array of security promises, especially if we extend the definition of security to mean states of being beyond "safety." What the state does do very well, and I show why and how in the course of this project, is reduce the complexity inherent in experiences bearing historical specificity to a "single locus"[65] and then proceed to contain said danger. If we, however, realize that every subject is at the same time a victim, perpetrator, bystander, and witness, then containing dichotomies and taming "the other" is far removed from the diagnosis that would eradicate the unforeseeable, the unknown, the dangerous. In order to illustrate this point more clearly, Peter Andreas contends that "border policing is a ritualistic performance. When the failures of the deterrence effort lead to a performative crisis, the performers save face by promising a bigger and better show."[66] The case of survivors of trafficking in human beings offers one such example of "rehabilitative" state power that inscribes itself as "the law" even at the moment when law bears witness to the inability of the nation-state to secure its subjects. So potent is the self-validation that ensues[67] that it is imperative to recognize and critique the seemingly all-accommodating nature of a state outside the law defined by a tension that, on the one hand, "institutes and makes" and, on the other, "deactivates and deposes,[68] all the while continuously seeking to affirm itself as the norm, despite the tension that defines it.

In the context of this project, insecurity is understood as constitutive of risk and not the other way around. The two are not interchangeable, though they inform each other and are often paired by the sovereign center, validated in continuing efforts to eradicate, stamp out, and fight terrorism and other dangers. This project does not interrogate the risks that terrorism poses to the security of nation-states and populations, a risk that exceeds the capacities of governments and societies to deal with it. The risks engaged with here are defined, understood, and tackled as bureaucratic problems that, when subjected to the careful eye and expertise of those in charge, are constructed as obstacles that can be minimized and/or altogether resolved in a desirable manner. In other words, the remnants of lived dangers inscribed on the bodies of human beings inform this work. Risk has become the prevalent modality enabling governing today, evolving beyond mere practices of governmentality and beyond Giorgio Agamben's exception as the state of governing outside the normal law. Rather, risk has come to dominate the modern nation-state imagination. It initiates and inaugurates practices of control and management that align but also exceed the ways in which states usurp power for themselves at the expense of their subjects' rights and freedoms.[69] The examples of trafficking survivors, refugees, and children fall under the category of "politicized identities" and, as such, "insofar as they are premised on exclusion from a universal ideal, require that idea, as well as their exclusion from it, for their own continuing existence as identities."[70]

While I subscribe to Wendy Brown's insight regarding the relationship between blame, *ressentiment*, and the excluded position many modern liberal subjects/groups occupy, the examples in this study point to something else and altogether just as powerful: the resentment toward "the other" that the citizen-subject experiences and that locks said subject into a relationship of dependency on sovereign power that can be read vertically to extend to the nation-state itself. The nation-state, caught in a network of insecurity it cannot appropriately address or even often identify, generates its raison d'être on the backs of excluded humanity. Important here is the distinction between excluded humanity and all citizens, the latter figuring as the recipient of security measures at the expense of their less fortunate liminal fellow humans, or aliens. As a result, violence and discrimination are normalized. Minorities are politicized not because of their ability to protest and challenge the status quo but for the danger inherent in their indigent status as such. The state imprints itself on said minority identities not as a protector but as an aggressor validating decisions regarding who's in and who's out. In this sense, the inherent state of anxiety that has come to characterize the modern subject demands that states wall people off to keep outsiders outside and insiders inside, irrespective of the fact that most threats are home grown.[71]

Perhaps the easiest starting point for a discussion of the ways in which risk as an organizing principle of modern governance has come to dominate the ways in which "power" and "security" are understood and constructed is Giorgio Agamben and his discussion of "the exception." Agamben is well known for his writings on the relationship between power, law, and the subject. In *Homo Sacer: Sovereign Power and Bare Life*, he asks, "What is the relation between politics and life, if life presents itself as what it included by means of an exclusion?"[72] For Agamben, the figure of *homo sacer* stands for life that can be killed but not sacrificed, depriving it of the power of law to protect it and of the power of the populace to mourn it. *Homo sacer* thus becomes doubly displaced from politics as we know it and instantiates a "new biopolitical body of humanity"[73] in which bare/natural life (*zoe*) and the life of the citizen (*bios*) overlap.

I suggest that Agamben's juridico-political idiom is not only unable to think through and accommodate the real-life implications of living with/at risk, but that it also represents a totalizing claim on truth and truth telling derived from an operationalization of dichotomies. Thus, *sacred life* as "presupposed and abandoned by the law in a state of exception" can be distinguished from *naked life* "abandoned to a kind of violence that is all the more effective for being anonymous and quotidian."[74] A definition of the human being as *homo sacer*, while informative of the experience of being a depoliticized subject during the Holocaust, for example, fails to address two important consequences of practices of politics of risk in the everyday.

It does not challenge the assumed calculability of the human as a product of a decision made by law, nor does it unearth modalities of engagement with individual experiences that are predicated and constructed on a belief in the totality of knowledge and sovereign power. The human as a "limit figure" predicated on a negatively constituted relationship to the juridical represents, on the one hand, "the radical crisis of every possibility of clearly distinguishing between membership and inclusion, between what is outside and what is inside, between exception and rule,"[75] while at the same time, s/he is recruited to speak on behalf of the sacred "insofar as it is taken into the sovereign exception."[76] If "only bare life is authentically political,"[77] then *bare life* either includes all of humanity or is specifically reserved as an identifying category for some and not other individuals. Refugees, prisoners, detainees, and the homeless, as carriers of *bare life*, are reduced to the violence that produces them, making citizenship something "one had to prove oneself worthy and which could therefore always be called into question."[78]

In the scope of this book, refugees are addressed beyond "limit figures." The political relevance of being a refugee is theorized as irreducible to the latter's inclusive or exclusive relationship to a sovereign center. The question of securing the body of the refugee is addressed not so much through the workings of a legitimately appointed sovereign apparatus but in terms of actual examples of out-living insecurity. Survivors of trafficking in human beings complicate and help clarify Agamben's position even further. Agamben's "inclusive exclusion of *zoe* in the polis"[79] should be read as suggesting both that politics were the place where life had to transform itself into good life and also the place where what had to be politicized has always already been bare life. Beyond security narratives, presented here are the stories of survivors of trauma at the hands of nation-states.

NOTES

1. Achille Mbembé, "Necropolitics," *Public Culture* 15, no. 1 (2003): 13.
2. Ibid., 14; emphasis in original.
3. Graham Burchell, Colin Gordon, and Peter Miller, eds., *The Foucault Effect: Studies in Governmentality* (Chicago: University of Chicago Press, 1991), ix.
4. Michel Foucault, "The Subject and Power," *Critical Inquiry* 8, no. 4 (1982): 777–95.
5. Daniel McLoughlin, "Giorgio Agamben on Security, Government and the Crisis of Law," *Griffith Law Review* 21, no. 3 (2012): 690.
6. Colin Gordon, "Government Rationality: An Introduction," in *The Foucault Effect*, 5.
7. Jackson Ackleson, "Constructing Security on the US-Mexico Border," *Political Geography* 24, no. 2 (2005): 165–84; Benjamin J. Muller, *Security, Risk and the Biometric State* (New York: Routledge, 2010).
8. Ulrich Beck, *World at Risk* (Cambridge: Polity, 2009).
9. Loise Amoore and Marieke de Goede, *Risk and the War on Terror* (New York: Routledge, 2008); Muller, *Security, Risk and the Biometric State*.
10. Roxanne Lynn Doty, "States of Exception on the Mexico-US Border: Security, 'Decisions,' and Civilian Border Patrol," *International Political Sociology* 1, no. 2 (2007): 113–37; Erik Camayd-Freixas, "Interpreting after the Largest ICE Raid in US History: A Personal Account," *Latino Studies* 7, no. 1 (2009): 123–39.
11. Étienne Balibar, *Politics and the Other Scene* (London: Verso, 2002), 29.
12. Ranjana Khanna, "Disposability," *Differences* 20, no. 1 (2009): 181–98.
13. Gordon, "Government Rationality: An Introduction," 14.
14. Ibid., 16.
15. Khanna, "Disposability," 193.
16. Ibid., 188.
17. Balibar, *Politics and the Other Scene*, 94.
18. Ibid., 65.
19. McLoughlin, "Giorgio Agamben on Security, Government and the Crisis of Law," 696.

20. Ulrich Raulff, "An Interview with Giorgio Agamben," *German Law Journal* 5, no. 5 (2004): 611.
21. McLoughlin, "Giorgio Agamben on Security, Government and the Crisis of Law," 703.
22. François Debrix and Alexander Barder, "Nothing to Fear but Fear," *International Political Sociology* 3, no. 4 (2009): 412.
23. Charlotte Epstein, "Guilty Bodies, Productive Bodies, Destructive Bodies," *International Political Sociology* 1, no. 2 (2007): 151.
24. Beck, *World at Risk*, 53.
25. Epstein, "Guilty Bodies, Productive Bodies, Destructive Bodies," 153.
26. Claudia Aradau and Rens van Munster, "Governing Terrorism through Risk," *European Journal of International Relations* 13, no. 1 (2007): 96.
27. Ibid., 98.
28. Ibid., 95.
29. Debrix and Barder, "Nothing to Fear but Fear," 411.
30. Michael Dillon, "Poststructuralism, Complexity and Poetics," *Theory, Culture and Society* 17, no. 5 (2000): 14.
31. Ibid.
32. Michel Foucault, *Herculine Barbin* (New York: Vintage, 1980), 187.
33. Dillon, "Poststructuralism, Complexity and Poetics," 15.
34. Wendy Brown, *States of Injury* (Princeton, NJ: Princeton University Press, 1995).
35. Griselda Pollock, "Liquid Modernity and Cultural Analysis," *Theory, Culture and Society* 24 (2007): 115.
36. Susie Orbach, "People in Distress," in *The Politics of Risk Society*, ed. Jane Franklin (Oxford: Polity Press, 1998), 96.
37. Louise Amoore, *The Politics of Possibility* (Durham, NC: Duke University Press, 2013), 6.
38. Roxanne Lynn Doty, *Anti-Immigrantism in Western Democracies* (New York: Routledge, 2003), 9.
39. Carl Schmitt, *Political Theology: Four Chapters on the Concept of Sovereignty* (Chicago: University of Chicago Press, 2006).
40. Françoise Vergès, "The Age of Love," *Transformation* 47 (2001): 7, 8.
41. Ibid.
42. Balibar, *Politics and the Other Scene*, 57; emphasis in original.
43. Brown, *States of Injury*, 175.
44. Linda Bosniak, *The Citizen and the Alien: Dilemmas of Contemporary Membership* (Princeton, NJ: Princeton University Press, 2008), 12.
45. Muller, *Security, Risk and the Biometric State*, 63–68.
46. Brown, *States of Injury*, 40–41.
47. Michel Foucault, *Discipline and Punish: The Birth of the Prison* (New York: Vintage Books, 1995), 42.
48. Balibar, *Politics and the Other Scene*, 87; emphasis in original.
49. Khanna, "Disposability," 185.
50. Bosniak, *The Citizen and the Alien*, 7.
51. Balibar, *Politics and the Other Scene*, 4.
52. Bertrand Ogilvie, *L'Homme Jetable* (Paris: Éditions Amsterdam, 2012).
53. Brown, *States of Injury*, 27.
54. Balibar, *Politics and the Other Scene*, 139.
55. Brown, *States of Injury*, 27.
56. Ibid., 62.
57. Foucault, *Discipline and Punish.*
58. Mbembé, "Necropolitics," 12.
59. Muller, *Security, Risk and the Biometric State*.
60. Giorgio Agamben, *Means Without End* (Minneapolis: University of Minnesota Press, 2010), 44.

61. Giorgio Agamben, *Homo Sacer: Sovereign Power and Bare Life* (Stanford, CA: Stanford University Press, 1998); Giorgio Agamben, *The State of Exception* (Chicago: University of Chicago Press, 2005).

62. Beck, *World at Risk*, 2009.

63. Antony Giddens, *The Consequences of Modernity* (Stanford, CA: Stanford University Press, 2011); Beck, *World at Risk*; Brown, *States of Injury*.

64. Beck, *World at Risk*, 41.

65. Debrati Sanyal, "Poetry in a Time of Terror," *Yale French Studies* 175/176 (2014): 105.

66. Peter Andreas, *Border Games: Policing the U.S.-Mexico Divide* (Ithaca, NY: Cornell University Press, 2009); Wendy Brown, *Walled States: Waning Sovereignty* (London: Zone Books, 2010), 91.

67. Ibid.

68. Brown, *States of Injury*, 87.

69. Ibid.

70. Ibid., 65.

71. Brown, *Walled States*, 70.

72. Agamben, *Homo Sacer*, 7.

73. Ibid., 8.

74. Ibid., 113.

75. Ibid., 25.

76. Ibid., 85.

77. Ibid., 106.

78. Ibid., 132.

79. Ibid., 7.

Chapter Two

The Spectre That Haunts

Giorgio Agamben

Giorgio Agamben's *homo sacer* as the being that can be killed but not sacrificed "and as such can be eliminated without punishment"[1] validates the statement: "Bare life is no longer confined to a particular place or a definite category. It now . . . dwells in the biopolitical body of every living being."[2] The natural life of the species (*zoe*) is the initial vehicle for Agamben's theory of the exception. *Zoe*, reconfigured as a stable, ethically stripped category signed over to the authority of state power over politicized life (*bios*), makes it possible for Agamben to move somewhat unproblematically between dichotomies such as "exception" and "rule," "victim" and "survivor," "man" and "nonman." "Exiled bare life, like the life of the camp inhabitant, the refugee, the bandit . . . is without a voice in public affairs and may be killed or kept alive without ceremony and without criminality."[3] While the creation of the new juridico-political category of enemy combatants, for example, is an example of the actual and uncensored recruitment of life by the political exception, I would like to suggest that there is a danger of oversimplifying the relationship between the exception, law, the human subject, and the ways in which these figures become known through practices of bearing witness.

If the "state of exception" has become the normalized condition of politics, then *homo sacer* turns into a juridico-political category of identification all its own: immunized against opposition and abstracted and estranged from practices of justice. Fixed in that way, *homo sacer* serves rather than challenges the sovereign by underwriting the latter's claims to unlimited power. Once derived as the product of the relationship between law in a state of emergency and the sovereign's will to the usurpation of power, *homo sacer* easily turns into a limit figure founded on the juridical vacuum of the "beyond the law." Alongside the rather monolithic picture of the concentration camp that Agamben presents in his *Remnants of Auschwitz*, the figure of *homo sacer* as the *Muselmann*, rather than problematizing the relationship between the sovereign and power, only undermines the latter's role in accounting for and justifying violence. What is more, the human being as *homo sacer* becomes an exclusive political problem posited against only a few ethical demands and unproblematically committed to the care of a monolithic institution: the sovereign. Once life becomes the referent object of politics as *homo sacer*, modernity is crowned as the most politically charged of all preceding eras.[4] By implicating the figure of *zoe* in the workings of the sovereign state, Agamben is able to reformulate Foucault's biopolitics vis-à-vis the "state of exception" and claim that the "production of biopolitical body is the original activity of sovereign power."[5] If we understood politics with the help of a theory of the limit, then positing the relationship between the human and politics becomes predicated not on a theory of committed relation but on a reality in which, once the state of exception overflows its boundaries to coincide with the normal order, everything becomes possible.[6]

The witness carries *natural life* beyond *bios*. For example, the concentration camp survivor as *zoe* is separate from his essence as a speaking being, which makes the passage between silence and witness an impasse within the limits of Agamben's theory. If politics is concerned with the human, and if human life exceeds the meaning that can be contained within a theory of the exception or the limit, then politics itself has to be thought differently. Far from oversimplifying Agamben, his insistence on maintaining dichotomous categories of identification as reference points for criticizing the precursors to those very categories invites further careful reading of the politically informed implications his writing has for a study of bearing witness and trauma. In the context of the War on Terror, for example, and much like Agamben's engagement with Holocaust survivors, "terror is currently instrumentalized to designate a collective original wounding . . . but it is also cultivated as a collective dread and anticipation of what is to come, and thus mobilized to legitimize a politics of perpetual war."[7] Additionally, this project adds to discussions on risk by showing the myriad ways in which insecurity is created, addressed, and perpetuated, alongside measures to justify the "construction of zones of precariousness and death around the globe."[8]

In his *Remnants of Auschwitz*, Agamben addresses the nostalgic and tragic figure of the witness as victim, suffering and rejected, inscribed in language as the always absent, impossible trace of (the dead) witness. Left to the philosopher, the former's testimony does not recover a voice as much as it mourns the subject's severed relationship to the law and the former's subsequent reduction to an object, a "nonman." Reverse engineering this relationship, being protected by the law becomes symptomatic of security, of a justified claim to (human) rights and freedoms, while being outside the law is presented as rejection, abandonment, and insecurity. "And law, in suspending the concrete custom and usage of individuals, has been able to isolate something like a norm,"[9] so the representation of the human continues to be constructed with reference to a demand to be fragmentary. The relationship of the subject to law in a "state of exception" is explained as a reversal of the norm that typically defines the sovereign's obligation to its subjects. Agamben seems to exploit the ambiguity inherent in the productive relationship that defines the interaction between the state, its subjects, and the law by cashing in on the idea that the concept of a human being contains within it the potential for a biopolitical fissure. "In every case, the state of exception marks a threshold at which logic and praxis blur with each other and a pure violence without logos claims to realize an enunciation without any real reference."[10]

The case of survivors of trafficking in human beings offers a pertinent example of such administration of life, in which government apparatuses are employed in the service of saving and managing but also improving and optimizing the lives of survivors. In this sense, and because risk and insecurity lurk ever present on the horizon, the nation-state can imply itself and its services in the workings of our daily lives, making "the exception," usually understood as a state of affairs in an emergency, a sufficient but not a necessary condition for the deployment of the political technology of what I call "emergent governmentality." I have chosen this term for three reasons: first, it stays close to Agamben's idea of "the state of emergency," yet steps away from it enough to allow us to think beyond Agamben's logic of exceptionality. Second, it points to a state in which governing as management is perpetually informed by a process of "emerging" and "evolving" risks, yet ones that do not simply underwrite an exception but also challenge and exceed it. Last, speaking of "emergent governmentality" allows for the discussion of various processes of government and points to the fluid, contingent, and undecided nature of sovereign practices.

Agamben's theory of the exception posits the state as unitary, sovereign, and rational when in fact the state is composed of a variety of institutions that govern together but also differently.[11] In this sense, it is important to posit states of exception such as Guantánamo Bay, Abu Ghraib, and refugee and detention camps as examples participating and defining a security dialogue introduced by a sovereign state first, in order to ensure its continued and varied use of power and only then because of threats and the fear of terrorist violence. Peter Passavant's summary of Scheppele's analysis of U.S. executive laws/measures post-9/11 suggests that "this emergence (and ruin) is hastened by those who seek to enhance surveillance and presidential powers, while diminishing the power of courts and legislative oversights as a response to September 11, 2001."[12] What this makes evident is the purposeful funneling of power in one branch of the government and/or in the hands of a single individual in an effort to legitimize, legalize, and normalize the very mechanisms otherwise antithetical to democracy. Thus, torture of enemy combatants, because they are not prisoners of war and thus unprotected under the Geneva Convention, ends up underwriting security and the law through violence.[13]

Once the "state of exception" becomes its own category of identification by claiming for itself a spatial, juridical (though defined through a lack), and human resource allotment, *homo sacer* comes to signify a way of life rather than a challenge to a way of nonlife. It is through such normalization that Agamben (using the examples of the prison, the refugee camp, and airport transit areas) suggests the linear nature of the subject's relationship to language (power). In the words of Peter Passavant, "In the state of exception law is in force even if specific legal prescriptions or prohibitions are suspended. When law is in force without significance, law is an empty potentiality that is so much in force without content that it becomes, as it did in the camps, indistinguishable from life."[14] Agamben defines pure potentiality as the potential to be and not to be at the same time. If this were true, then the "state of exception" read as pure potentiality cannot be defined entirely either in positive or negative terms but must be seen rather as "double possibility."[15] What this comments on is the productive potential contained within every possible decision, the potential to decide against deciding, the potential to be and not to be. In other words, "Agamben's work enables us to analyze what is at stake in the politics of the decision. He elaborates on how sovereign power operates through the state of emergency and how the very positing of the question through the trope of emergency is always already on the side of sovereignty."[16]

In the context of this book, refugees, trafficking survivors, and children victims of war can be read as limit figures but also as more than that. This tension sets up important roadblocks to our ability to unproblematically assign blame for suffering or victory to any one agent or event and forces us to accept the multiplicity of slippages that define insecurity as "a field of testimony and engagement that enables a critique of terror's contemporary articulations."[17] Allow me to suggest a preliminary hypothesis for my discussion of the limit: if, following from Agamben's discussion of *potestas*, we agree that being in the condition of exception means not to remain in that condition, then *homo sacer* as a limit figure deconstructs the unproblematic positing of the "state of exception" as norm. "What emerges in this limit figure is the radical crisis of every possibility of clearly distinguishing between membership and inclusion, between what is outside and what is inside, between exception and rule."[18] The inability to distinguish clearly between these binary categories, even as they are instructive of the juridico-political nature of the sovereign exception, threatens to universalize our thinking of the subject (as *bare life*) on behalf of the "exceptional" space of the political. Thus, in Agamben's world, "one mode of existence is, territorially predominant, in being elevated or affirmed and another denied. There is exclusion here, but also inclusion. And for law, inclusion is irreducible."[19] In this sense, the example of survivors of trafficking in human beings showcases the ambiguous position they occupy vis-à-vis the state as victims. If "inclusion always *exceeds* membership" and if "the exception expresses precisely this impossibility of a system's making inclusion coincide with membership, its reducing all its parts to unity,"[20] then is the singularity of *homo sacer* recognized through a movement of generalization that effaces it? Yes and no.

In *Means Without End*, Agamben is even more explicit when he engages the figure of the human. He defines *zoe* as natural life common to all living beings and *bios* as the manner of living peculiar to a single individual or group,[21] making clear the reciprocal relationship between power, rights, and the state thus: "In the system of the Nation-State, what are sacred and inalienable rights are conceived of as such only when they relate to the figure of the citizen, *stated* individual. The rights of *naked life* are not endorsed by the state."[22] The holy trinity of man-body, citizen-rights, state-power still permeates discourses in international relations despite the ambiguities underwriting "the strategic rationality between these three."[23] Naked life, defined in its exclusion from the national order, is driven outside the bounds of state protective power, underwriting the very exception that keeps it confined to a state of exclusion. "Sacred life," in turn, becomes "life that is presupposed and abandoned by the law in a state of exception—it is the mute carrier of sovereignty, the real sovereign subject," whereas naked life is "abandoned to a kind of violence that is all the most effective for being anonymous and quotidian."[24] The managing techniques of governmentality exceed sacred life but do not unproblematically stand for naked life either. In the space between manifestations of sacred life as the life of the homeless citizen and manifestations of naked life as, for example, the life of the deported illegal migrant or the inmate at Guantánamo Bay, figures the life of subjects whose claim to belonging is validated in the process of validating and giving free, willing consent to the apparatuses of the nation-state to inscribe its mechanisms of governmentality on its body. The example of survivors of trafficking in human beings, as they willingly inscribe themselves into the powerful machine of state governmental workings, demands special engagement away from both Foucault and Agamben.

Despite Agamben's insightful discussion regarding the legal channels through which the law itself validates its prohibition on the ability of the state to break it, the book argues that Agamben's focus on the juridical nature of the exception, while hugely informative, is also limiting. It affects our ability to think through the ways in which a carrier of sacred humanity holds a political charge able to problematize, question, and unsettle the sovereign order. Michel Foucault is much more helpful here, as he addresses precisely this relationship in *The Subject and Power*, writing resistance into the body of the free subject and almost precluding inaction as a choice, especially in the context of a society in which security has "become the dominant component of modern governmental rationality."[25] The *state of exception*, itself a construction and somewhat of an exclusively juridical problematique, will be interrogated in relation to the question of making life live in ways that address the relationship between the *norm* and the *exception* practiced and negotiated every day beyond sovereignty. The example of Tibetan refugees in Nepal and their precarious legal situation, or rather their lack of legal status, showcases opportunities for negotiating prohibition through a recourse both to the semi-state-like entity of the Tibetan Government in Exile and to ideas of home interpreted as "being together."

In fact, the exclusion become norm, as Agamben argues, offers, if not an overhaul, then a clear entryway into an engagement with state lawlessness, violence, and exclusion. While the exception may not be *the* condition for the ontological undoing of the nation-state, we can still engage the epistemological markers that uncover the bodies of the citizen, the illegal, the migrant, the trafficking survivor, the diseased or hurt human as *homo sacer*. To reiterate with reference to Agamben's own words, "The state of exception therefore appears as an 'illegal' but perfectly 'juridical and constitutional' measure that is realized in the production of new norms."[26] In the context of modern political power, no longer is the exception rooted in the figure of a dictator. The latter is superfluous to the operation of exceptional power situated outside the register of hard, tyrannical, or oppressive power in a "world risk society" in which prevailing risks are recognized but impossible to alleviate and address satisfactorily.

If Agamben's *state of exception* has become, as he wants us to believe, the underlying characteristic of modern life, and if the juridico-political suspension of law(s) has assumed real life dimensions, then what becomes of bearing witness to the *exception*? What could possibly be the truth-affirming power of a witness that informs of what it means to be human through the lens of inhuman, intolerable suffering? Perhaps Agamben's question, "What is the relation between politics and life, if life presents itself as what is included by means of an exclusion?"[27] ought to be read not so much as positing the figure of *homo sacer* but as an invitation to understand "bare life" itself as a state of being human that problematizes the project of the exception itself. In that sense, (human) life becomes predicated beyond the juridico-political relationship amalgamating both *zoe* and *bios* into a *homo sacer* and rather as a state of being informed by an irreducible, ethically constituted relationship to *natural life* as *zoe*. While the relationship between power, the law, and the subject (in a "state of exception" or not) is central to being able to establish and define a criteria for attributing blame and restoring justice, a discussion of bearing witness cannot limit itself to the juridico-political significance of testimony that, recorded as part of a discourse informed by terms such as "victim" and "perpetrator," "guilty" and "innocent," "man" and "nonman," fails to account for those experiences that are not as easily addressed by the juridical language of "exception."

Agamben's *homo sacer*, though highly informative of the precarious and often ambivalent relationship between state-sponsored violence and the law, is a term of identification that cannot unproblematically be applied to all situations falling under the category of juridical emergency. A "state of exception" is what necessitates, nurtures, and gives meaning to the term *homo sacer* and that subsequently validates and fills in the legal blank that is the subject/citizen in a theory of the "exception." While the being that can be killed but not sacrificed might be an adequate term applicable to those who perished in the concentration, detention, and work camps of the Holocaust in its essence as borne out of a historically specific event (the Shoah), *homo sacer* cannot be referenced just as liberally to other instances in which the law, water, food, or shelter, for that matter, have been suspended or given over to the monopoly of a sovereign center.

In the context of this book, Agamben does make for a logical theoretical choice because he speaks so much of those excluded or banned from the sovereign sphere and precisely because the focus of this project is partially on people who are being readmitted, or in the process of being readmitted, as full-fledged members of the sovereign sphere. Survivors of trafficking in human beings constitute just one such example, as do long-term refugees, though their inclusion is often a consequence of lifelong state of exile. It is this process of readmittance that offers a glimpse at the mechanisms and technologies employed by the state in signifying its active engagement with "securing" the lives of its subjects and, in the case of children, owning their death. While Agamben holds true saying that "what confronts us today is a life that as such is exposed to a violence without precedent precisely in the most profane and banal ways,"[28] the life of the survivor of trafficking in human beings, for example, technically does not qualify as *homo sacer*, benefitting as it were from the machinery of the state in its status as victimized being in search of redress, yet the figure of the survivor is also an exceptional being, existing outside the norm of productive, lawful, disciplined citizenry. So how is Agamben's "exception" refined, challenged, and tested by instances of state intervention and investment into the lives of what I see as quasi–*homines sacri* who don't fully fit the bill of humans living the exception, yet who are linguistically (in their status as trafficking survivors), juridically (implied in the force of the law to rectify the violence done to them), economically (often at the mercy of state handouts for up to a year), and politically (inscribed in the state order only after an admission of victimhood instrumental to the ability of nation-states to exercise power over rejection and shedding of said victimhood) positioned in the registers of life that, if not eliminated, can be abandoned with ease?

The point of contention here is not so much that the legal suspension of law by itself does not belong to and define the "state of exception" but rather that *homo sacer* as a category of identification read as "the bare life of the citizen" has been defined as "the new biopolitical body of humanity."[29] The figure of the concentration camp inmate, that is, *bare life* within a now normalized "state of exception," does little to challenge a reading of politics as a periodical taking turns of normal and abnormal periods. If anything, it underwrites an understanding of the state as a monolithic, unified, and rational actor whose relationship to the exception is constituted by a parsimonious theory of the exception. Nowhere does Agamben problematize the monolithically posited nature of the categories of "norm" and "exception," formally engaged as each other's opposites and often used as tools of problematizing other fixed categories. The "state of exception" disguises practices of power abuse already inscribed in the normal workings of state(s) as instances of emergent security but also as state insecurity. Mechanisms of securing the state should also be read and understood as methods for perpetuating insecurity in people and unveiling the insecurity of the sovereign center. This, in turn, leads to a state in which the "exception," now tracked and modified exclusively to the sovereign center, comes to stand for the law in a relationship that sooner or later grants legitimacy to the "exception" and inaugurates it as the new "norm." The political, made subject to such objectives, is thought solely as security. Thinking the political differently, as a desire to out-live security,[30] demands a language/a toolbox/an opening that speaks on behalf of this insecurity. The cases presented in this book attempt to do just that.

Out of this new "norm" another exception is born in due time and so on and so on, which all represent instances of control on the part of the sovereign that, in the latter's attempt to normalize and discipline his/her subjects, underwrite the everyday workings and dissemination of power. Within Agamben's argument, as far as the figure of the human being is concerned, s/he ends up being theorized on the margins that validate both the "norm" and the "exception," the two exercising upon it as mechanisms of security. The actual needs that the human faces are unaccounted for by Agamben, who would rather call for the termination of the state (in favor of a reality of "pure potentiality") than recognize the actual potential state institutions have to ensure, protect, and perpetuate justice, democracy, and equality. While Agamben's movement between the "norm" and the "exception" is clear enough theoretically, an engagement with the terms of "norm" and "exception" as such creates its own lacuna within the everyday life of human beings, engaged with in this project through the figures of trafficking survivors, refugees, and children whose lives were lost as a result of violent conflict. How can this lacuna be filled? Should it be filled?

As Michel Foucault has clearly shown, it is not necessarily the case that what is called the "norm" is, as such, normal, or that there has ever been a political order whose organization and rules were exempt from the workings of power, or, for that matter, that such an order is ever achievable.[31] Veena Das cautions against the unproblematic recalling of painful memories in language thus: "even the idea that we should recover the narratives of violence becomes problematic when we realize that such narratives cannot be told unless we see the relation between pain and language that a culture has evolved."[32] Das asserts that ideas become institutionalized by certain historical, societal, and political practices and come to be understood only from within very inflexible boundaries. The search for flexibility informs the use of poetry as a discursive tool throughout this project, as it is my contention that poetry is best able to accommodate the "infinite mourning for the departed"[33] that instances of suffering and trauma as a result of insecurity bring about.

In the sense in which the state (as sovereign, law-deciding, policing, legitimating actor) can exercise power in constructing the kind of truth that is made public, the practice of bearing witness becomes directly implicated in the process of power dissemination and resistance to said power. State discourses elicit meaning, but so do discourses of survivors of trafficking. An officially recognized (theorized) *state of exception* fails to recognize the practices through which power becomes infiltrated and mutates, especially through everyday language. This instance of bearing witness through the idiom of everyday speech is privileged as the site of truth making. On the contrary, the *state of exception*, in the moment it recognizes and proclaims itself as the norm, enters into a relationship with power and truth telling that abandons itself to the presumed autonomy and unity of the sovereign. As Jenny Edkins has suggested, "the testimony of survivors can challenge structures of power and authority"[34] that reproduce a certain social order. In a similar fashion she has suggested that, while helpful in certain instances, linear narratives end up depoliticizing and sterilizing trauma. Her call for narratives "encircling the trauma" alerts to an interesting possibility. Bearing witness, predicated on the ability to make the past known, exposes not only the workings of sovereign power vis-à-vis language but also the ability of the witness to construct, reshape, and manipulate truth. A problematization of witnessing as truth telling alerts not only to the ways in which truth can be suppressed, rejected, and constructed but also to the ways in which truth telling necessarily anchors the witness in a relationship dependent on the sovereign. A closer look at the witness as victim and survivor, as subaltern and marginalized human being, draws attention to processes of truth telling within which the witness constructs her/himself once, vis-à-vis the perpetrator, then the judge as representative of institution(s) of reparation, then again history as written in the archive, and finally the need to construct a "first-hand" narrative in the public sphere that will remember, pass on, recognize, and grant the witness-as-survivor his/her due place in cosmologies of sympathy, healing, and recognition. Examples of such narratives in this book point to how this is done and what its effects are.

This move is important because a positing of *homo sacer* as a limit figure makes the latter into a spectacle. Much like the process of bearing witness, the status of a survivor is not a question of all or nothing, of an exception or a norm, of language or silence, of truth or falsity. The condition of undecidability that necessitates a constant interrogation and exchange between dichotomous categories within a juridico-political register underwrites a state of affairs in which opposites, in order that they be possible, must oscillate between "a rational mode and a magical mode of being."[35] The realm of law becomes illegible, shot through with ambivalence, referenced upon the exclusions that make possible its regulatory character.

The *state of exception* is not simply a phenomenon that appears suddenly, alters the "normal" ways of doing business, and then, once it has run its course, gives way to a new normal way of being until another emergency sweeps through the registers of the juridical. It is predicated on and enables the permanent creation of liminal beings, unwanted subjects, and illegal aliens, to name a few, whose insecurity breeds not only deprivation, but also novel forms of resistance. For example, illegal immigrants have to negotiate their ability to remain in a country and work, a negotiation that is not always friendly or traditional. In the cases of immigrants at Woomera Detention Centre in Australia, sewing one's lips evolved into protest through a performance of a corporeal politics of self-mutilation. Migrants tearing up their papers to escape notice and "categorization" by the nation-state is another example of resisting imposed rightlessness.[36] To even posit a *state of exception* as norm means to posit a world in which power's hiding places become predicated upon the continuous (re)production of limit figures necessarily mediated by processes of exclusion and opposition, filtered always through the prism of the juridical that is beyond the human, exceeding but also erasing it.

How do manufactured uncertainties and not-knowing, after Ulrich Beck, fit within the contemporary global imaginary of nation-state power? As uncertainties exceed efforts at mitigating their negative consequences, how are they tackled by sovereign power obsessed with the impression of providing security? If experts are failing at managing risks, and if governments find themselves fundamentally unable to and continuously failing at addressing environmental, terrorist, and other threats, what is to suggest that they will miraculously get themselves in gear? Last, if the "traumatic experience of the enforced community of global risks that threaten everyone's existence"[37] is what introduces a sense of equality, how does this reckon with a reality of "radical individualization" called upon in response to "authoritarian failed states" and their inability to secure the individual? Perhaps Ulrich Beck is correct that "the bitter irony here is that manufactured uncertainty (knowledge), insecurity (welfare state) and the sense of insecurity (violence) undermine *and* confirm the power of the state beyond the scope of democratic legitimation,"[38] so much so that the nation-state cannot renounce/decry its recourse to the exception, seeing how the latter has come to dominate the legal subject's vision of itself and its rightful place in a universe of chosen, included, legal, recognized, and safe subjects as opposed to hordes of illegal, unrecognized, dangerous others.

As already stated, I believe the "exception," though informative, is an imperfect and misguided way of looking at what nation-states do every day, with their failure to provide security and prosperity to their citizens secondary to their desire and commitment to perform said security through various tools of governmentality. So while national governments continue to be held accountable for addressing threats, the latter have been transformed into a visual and optic illusion the construction of which, through proper advertising, border campaigns, special ops missions, and discursive eloquence, preempts performances of everyday security that outlive old-fashioned security. In the process, acts aimed at preempting security are performed daily and all around us, to the effect of legitimating their own claim to being, in fact, security producing.

This would explain why "the creation of legal identities, the creation of legal spaces and of legal borders, and the definition of the scope of legal rights"[39] holds so much power over the lives of both legal and illegal, righteous and criminal, domestic and alien individuals and why such measures of control "can be easily produced and reproduced without resorting to an exceptional tool of government."[40] The latter is largely underwritten by nation-state walls as "modern day temples housing the ghost of political sovereignty" conferring "magical protection against powers incomprehensibly large, corrosive, and humanly uncontrolled."[41] In fact, ordinary tools of government will be shown for their ineffectiveness at alleviating risk, especially when it comes to survivors of trafficking in human beings and refugees.

Pursuant to these realities, the book does not so much problematize risk as such, as its permeating presence on the modern political horizon is unavoidable. It also does not engage with the biometric realities of sovereign power and the pervasive ways in which the state has been creeping in the lives of us all through surveillance, increased travel restrictions, in our everyday lives as parents, travelers, consumers, employees, and recipients of government benefits of one kind or another. These are all questions that have been engaged with elsewhere by scholars better equipped to comment on them than myself.[42] The project, rather, engages with examples of "everyday security"—

namely, the realities and everyday lives of people whose production as "victims" of insecurity has necessitated that they negotiate living with risk long after they attempted, succeeded, or failed to cross an international border. While the demands of everyday existence for us all are pressing and worth engaging, the project contrasts state handling of "victims" with their own experience of making life possible. Navigating the trauma of exile, trafficking, or war becomes an important element in their recovery. This trauma and its effects on the subject are engaged throughout the project in creative ways, allowing for different avenues of listening to but also interpreting trauma.

The book attempts a slight reversal in terms of how we think about insecurity. Rather than conceive of individual identities and securities as products of various state mechanisms aimed at combating, averting, and mitigating risk, the example of victimized populations offers a glimpse at the ways in which national security has come to underwrite a space of "emergent governmentality," where insecurity no longer shocks or is even tackled beyond ever emerging and evolving mechanisms that create more insecurity and risk. In this sense, Michel Foucault's assertion that "fear is not actually something that the modern state and its agents ever want to do away with or be free from. Rather, fear must be produced and reproduced by governmental agents in order to establish control, supervision, or enhancement of the social body through multiple mechanisms of measurement, calculation, improvement, and preservation of life."[43] In this sense, and with reference to survivors of trafficking in human beings, for example, the fear that governs state logics justifying exclusion, lawlessness, and chaos is recast and repackaged as state preparedness, state mitigation exercises, and state risk and disaster management plans. In other words, risk becomes the new modus operandi of "emergent governmentality," and the subjects in need of securing carry out the important task of validating the myriad practices underwriting exercises in tackling, but never quite eradicating, fear and the unknown. The desired effect is not to remain insecure but to understand said insecurity as an opportunity to become more fully aware of and cognizant of one's role in defining said (in)security.

In the chapters that follow, this state of "permanent risk" will be engaged through examples of human beings caught in the mechanisms of state power and whose bodies on the move are governed, produced, and managed through discourses of pathologization. As a result, "A biopolitical production of fear implies that a multitude of agents *for* government are in the business of deploying a range of techniques, procedures, and tactics that strive to realize control over or regulation of the population's everyday conduct from all sorts of points, positions, or perspectives and through various relations/lines of force," leading to a widespread situation in which "security mechanisms or safety arrangements . . . are constantly at work mobilizing fear/terror."[44] Governing thus becomes an exercise in the production and maintenance of fear that works most pervasively and with greatest force on human beings on the move. An intended consequence of this reality is one anticipated by Ulrich Beck, in which national and global risks, products of modernity's march to progress, become largely individualized and thrown back on subjects expected not only to mitigate them but also to take responsibility for their inevitable personal failure in arresting their forward march.

Michael Dillon's "emergent life" is reminiscent here of the kinds of everyday realities that populations are faced with, though "emergent life" is not necessarily only the product of governing by terror per se but also the product of "emergent sovereignty."[45] In the context of this project, Dillon keeps Agamben's "exception" alive by providing an opportunity for survivors of trafficking in human beings, in addition to being disposable, to be understood as sites of critique of state practices, and where the subject/citizen/human in question is no longer simply "removed from the social compact per se but instead will remain as the main political modality of government."[46]

As a result, survivors of trafficking in human beings can only think of themselves in relation to the mechanisms governing their rehabilitation; refugees and other displaced humans think and conceive of themselves as they hopelessly, though sometimes also successfully, search for a validation by state apparatuses and mechanisms that may have been the very actors that denied their claims to belonging; where the dead bodies of children drowned in the waters of the Mediterranean and beyond enter memory in their role as quasi-subjects or, even better, as victims of nation-states outside the need for recognition. In an effort to move away from an unproblematic equation of disparate individuals (trafficking survivors, children, refugees, etc.) with each other, and cognizant of Wendy Brown's warning regarding the reification of the truth of "politicized identities,"[47] what follows is an attempt to honor the individual testimonies of people as I heard them while working out how their experiences might inform narratives of security within international relations.

NOTES

1. Giorgio Agamben, *Homo Sacer: Sovereign Power and Bare Life* (Stanford, CA: Stanford University Press, 1998), 139.
2. Ibid., 140.
3. Thomas Carl Wall, "Au Hasard," in *Politics, Metaphysics, and Death*, ed. Andrew Norris (Durham, NC: Duke University Press, 2005), 39–40.
4. Andrew Norris, "Introduction: Giorgio Agamben and the Politics of the Living Dead," in *Politics, Metaphysics, and Death*, 2.
5. Agamben, *Homo Sacer*, 6.
6. Paul A. Passavant, "The Contradictory State of Giorgio Agamben," *Political Theory* 35, no. 2 (2007): 156.
7. Debrati Sanyal, "Poetry in a Time of Terror," *Yale French Studies* 175/176 (2014): 103.

8. Ibid., 104.
9. Giorgio Agamben, *State of Exception* (Chicago: University of Chicago Press, 2005), 37.
10. Ibid., 40.
11. Passavant, "The Contradictory State of Giorgio Agamben," 169.
12. Ibid., 170.
13. Alex Bellamy, "No Pain, No Gain? Torture and Ethics in the War on Terror," *International Affairs* 82, no. 1 (2006): 121–48; Jamal Barnes, "The 'War on Terror' and the Battle for the Definition of Torture," *International Affairs* 30, no. 1 (2015): 102–24; Rodney Allen, "Torture, Criminality and the War on Terror," *Alternative Law Journal* 30, no. 5 (2005): 214–18; Julie Gerk Hernandez, "The Tortured Body, the Photograph, and the U.S. War on Terror," *CLCWeb: Comparative Literature and Culture* 9, no. 1 (2007): 2–10.
14. Passavant, "The Contradictory State of Giorgio Agamben," 156.
15. Ibid., 158.
16. Jenny Edkins, *Trauma and the Memory of Politics* (Cambridge: Cambridge University Press, 2003), 212.
17. Sanyal, "Poetry in a Time of Terror," 108.
18. Agamben, *Homo Sacer*, 25.
19. Peter Fitzpatrick, "Bare Sovereignty," in *Politics, Metaphysics, and Death*, 70.
20. Agamben, *Homo Sacer*, 70.
21. Giorgio Agamben, *Means Without End* (Minneapolis: University of Minnesota Press, 2010), 3.
22. Ibid., 20.
23. Ibid.
24. Ibid., 113.
25. Colin Gordon, "Government Rationality: An Introduction," in *The Foucault Effect*, ed. G. Burchell, C. Gordon, and P. Miller (Chicago: University of Chicago Press, 1991), 20.
26. Agamben, *State of Exception*, 28.
27. Agamben, *Homo Sacer*, 7.
28. Ibid., 114.
29. Ibid., 8.
30. I borrow the term "out-living" from Michael Dillon's *Politics of Security* (London: Routledge, 1996). A closer engagement with the bigger question of "out-living security" will be offered in the last chapter of this book.
31. Michel Foucault, "Governmentality," in *The Foucault Effect: Studies in Governmentality*, edited by G. Burchell, C. Gordon, and P. Miller (Chicago: University of Chicago Press, 1991); Michel Foucault, *Discipline and Punish: The Birth of the Prison* (New York: Vintage Books, 1995); Michel Foucault, *Power/Knowledge: Studies in Governmentality* (Chicago: University of Chicago Press, 1980).
32. Veena Das, *Life and Words: Violence and the Descent into the Ordinary* (Oakland, CA: University of California Press, 2006), 57.
33. Sanyal, "Poetry in a Time of Terror," 103.
34. Edkins, *Trauma and the Memory of Politics*, 5.
35. Das, *Life and Words*, 162.
36. Tugba Basaran, "Security, Law, Borders: Spaces of Exclusion," *International Political Sociology* 2 (2008): 339–54.
37. Ulrich Beck, *World at Risk* (Cambridge: Polity, 2009), 56.
38. Ibid., 79–80.
39. Basaran, "Security, Law, Borders: Spaces of Exclusion," 341.
40. Ibid.
41. Wendy Brown, *Walled States: Waning Sovereignty* (London: Zone Books, 2010), 133.
42. Loise Amoore and Marieke de Goede, *Risk and the War on Terror* (New York: Routledge, 2008); Benjamin J. Muller, *Security, Risk and the Biometric State* (New York: Routledge, 2010); Claudia Aradau and Rens van Munster, "Governing Terrorism through Risk," *European Journal of International Relations* 13, no. 1 (2007): 89–115.
43. François Debrix and Alexander Barder, "Nothing to Fear but Fear," *International Political Sociology* 3, no. 4 (2009): 400.

44. Ibid., 408.
45. Michael Dillon, “Governing Terror,” *International Political Sociology* 1 (2007): 7–28.
46. Debrix and Barder, “Nothing to Fear but Fear,” 411.
47. Wendy Brown, *States of Injury* (Princeton, NJ: Princeton University Press, 1995).

Chapter Three

Trafficking and the Single Story

MORE THAN MEETS THE EYE

This is a chapter about misunderstandings, misrepresentations, unexpected turns, aporias, and impasses, but also about humility, friendship, and courage. To write about trafficking in human beings is intimidating and scary, seeing how writing usually stands in for at least *some* commitment to truth telling and expertise. Because the majority of the original assumptions that drove this research were changed along the course of the research, what I was left with was a project about state power, the proliferation of risk, and the meaning of borders that refused to fit easily within my preselected academic parameters. Not only was I theoretically and disciplinarily at sea, but there were also difficulties in conducting field research due to the "gate keeping" that became part of daily negotiations for access to trafficking survivors—access that was defined against survivors' empowerment needs delivered by NGOs, governments, clinicians, and other well-wishers fully committed to survivors' safe keeping away from my prying recorder.

The impasses and aporias encountered along the way, in fact, did shed light on questions of state power, legitimacy, borders, and risk, but in a backward sort of way. When speaking with friends, family, and colleagues, the admission that I went to Bulgaria for six months to meet and work with survivors of trafficking in human beings usually led to a knowing nod of the head and exalted praise for my humanity, selflessness, courage, and empathy by all bystanders, so much so that I was truly convinced of my ability and, indeed, suitability to conduct novel, important, and truly life-changing research with people who somehow needed me to write about what I had found out about them. I am happy to report I failed at conducting novel, important, and life-changing research and shattered all my expectations and assumptions in ways I hope continue to rattle me, and the reader, for years to come.

Here is what this chapter is not about: it is not about the poor, innocent, uneducated, and exploited "victim" of trafficking, though it is about the lives of ordinary people who found themselves involved, implicated, and touched by the trafficking industry; it is not about the life-changing, heroic efforts of governments to save and rehabilitate these "victims" for theirs and the greater good, but it is about the intersection of administrative care, sovereign expectations, and international legal mechanisms and obligations that all make of antitrafficking initiatives quite the entrepreneurial undertaking. It is not about the proliferation of risk as the new, all-sweeping turn in political science, so much so that we all freeze mid-step, petrified by the dangers that lurk behind every corner, be they embodied by black men, Arabs, the poor, the rich, foreigners, or "others" in general. It is about the official management of vulnerability and the linguistic-juridical-sovereign machinations producing a certain regime of insecurity that is felt, lived, and negotiated with the goal of securitizing itself and being secured by others. The chapter is also about the power of storytelling, about visions of voice, long walks, early morning coffees, Turkish soup on a rainy day, French peasant cakes in a sidewalk café, chain smoking, getting lost, gift exchanges, long bus rides, friendships, and, of course, Gypsy dancing.

The chapter problematizes the conceptualization of trafficking, vulnerability, and risk as they inform each other and produce a persistent, legitimate, and authoritative discourse of insecurity *outside* the state. This insecurity is then tackled, handled, and mitigated by various *inside* powers. It investigates ways in which mainstream engagements with vulnerability and risk contribute to a static view of the trafficked human conceptualized vis-à-vis the state either in the context of needing saving or as a source of insecurity. The chapter suggests that one of the unexpected consequences of this process has been the casting of sovereign fragility as sovereign strength, all the while superseding, bypassing, and overshadowing instances of individual agency/autonomy as such—autonomy evident in stories of life after trafficking. In my interactions with trafficking survivors, I surmised a general sense of insecurity deriving not directly from some universal condition of human rights abuse/denial, but one grounded within the borders of *the* home state, the latter understood less in the social contract sense of a "protector" and "rights giver" and more in the sense of an insecurity-generating site of belonging by virtue of birth, language, accident, poverty, deportation, or geography. Trafficking in the Bulgarian context is the case study used to test assumptions about risk, vulnerability, agency, insecurity, and citizenship.

The "saving" of trafficking "victims" is examined in the context of creating and sustaining several hierarchies: a dependency by Bulgarian authorities on European Union (EU) directives and funding, a dependency by survivors of trafficking on state support and funding (both short term), a dependency by the EU on the cooperation of Bulgarian authorities, and last, a dependency by the Bulgarian government on the proliferation of human trafficking in order to secure recognition for its ameliorative actions aimed at combating the illegal transportation and exploitation of its citizens by its citizens and foreigners across international borders.

What interests me here is not so much finding out whether trafficking survivors live with/at risk or not but rather how both risk and agency challenge the assumption that either agency or risk are simply given or taken. Taking its lead from Vicki Squire's discussion on borders and migration in *Post/Humanitarian Border Politics Between Mexico and the US*, this chapter will showcase the ways in which the trafficking survivor is "produced" and made to epitomize risk and vulnerability in the process of interacting with and being managed by various humanitarian and governmental interventions.[1] The trafficked human is unproblematically and officiously equated with trafficked populations as such and often lost along the way. Because trafficking in itself constitutes a complicated and entangled network of relationships, it cannot be reduced unproblematically to an engagement with "the victims" alone. Yet and even though human trafficking in itself does not pose a direct threat to the state, it has been mobilized in service of definitions of national security, so much so that the latter, rather than allowing for wider ways of thinking about trafficking, has actually led to manufacturing and justifying wider ways of thinking about security. In this sense, the enactment of agency away from said practices of securitizing the state[2] requires that the insecurity plaguing the nation-state be challenged from the point of view of the material, corporeal essence of the individual within it, from the point of view of *the human in the body*.

The chapter unfolds as follows: first, I briefly go over trafficking definitions, some trafficking statistics, and offer a summary of the literature on trafficking by way of establishing a baseline against which to make my arguments. A brief discussion of the role and practices of various state and nongovernmental agents of humanitarian redress will illuminate the enactment of security on the "victim's" body and help introduce Giorgio Agamben and his thought on "the state of exception" and "bare life." Trafficking in the Bulgarian context is discussed as an example of a process of pathologization of trafficking and its "victims" but not the state. I argue that "trafficking as pathology" obfuscates the myriad ways in which state security is contingent on "the construction of women as risky beings" serving to "depoliticize their agency in resisting the *dispositif* of security."[3] As a result, acceptable behaviors and forms of agency are identified, constructed, and mitigated by the institutions of government. In this sense, trafficking becomes the fuel for a humanitarian machine composed of the United Nations, the EU, governments, social workers, and nongovernmental organizations. These organizations comprise the gargantuan apparatus engaged in remedying trafficking, yet an apparatus eventually displacing survivors and their testimonies in favor of directives, orders, conventions, and funding priorities. The two chapters that follow engage with the work of said apparatus parallel to engagements with the words and experiences of survivors. This is why throughout the course of the chapter the words of survivors of trafficking are presented as often as possible in poetic form. I trust the poetic form to encompass within it the lacunas and omissions that a linear narrative smooths over. Additionally, poetry precludes easy summary and export and thus forces one to engage with the "human in the body" and the fact that as a political subjectivity s/he counts and is thus not invisible.

The pathologization of trafficking, when examined closely and in conjunction with specific processes of risk creation/sustenance, reveals as false the assumption of modern governance that risk leads to insecurity, which necessitates the management of said risk by states, underwriting, as a result, claims to legitimate control over the choices, rights, and political nature of the modern subject. Rather, the chapter argues that vulnerability is not withdrawn but rather bestowed upon citizens/subjects by nation-states that is not an exceptional condition, after Giorgio Agamben, but a "consequence[s] of normal modes of government."[4] Thus, it is not a secondary and accidental consequence but an original and natural state inscribed on citizens and others alongside efforts to secure the whole against an intruding, abject, deformed, deviant "other,"[5] validating and normalizing democratic/neoliberal values and measures of control.

Ironically, rather than bulk under the weight of this reality, here lies the possibility for challenging, resisting, and rethinking sovereign security away from the exception. If we realize that the problem of security "is not its exceptionality, but rather its banality," where "the limitation of rights is in fact part of the common repertoire of liberal states,"[6] then perhaps we can garner tools away from discourses on exceptionality and thus better combat the myriad practices of being made insecure that ordinary people experience both as members of and outsiders to states. The example of survivors of trafficking in human beings is just one case in point and offers a place to begin interrogating and complicating the discourses of risk and insecurity governing our lives. I argue that the trafficked survivor belongs to a category of people who, alongside the refugee, uncover an inherent inability of nation-states to provide security for their citizens,[7] even as the former legitimize their claim to power and the exception through discourses and practices aimed at perpetuating freedom and rights through security. Trafficking underwrites the myth of state security: discussing the state as if the state were secure, discussing subjects as if subjects were enjoying liberty and equality, discussing risk as if there were a risk-free wonderland one could recover simply by creating more institutionalized safeguards, obscuring the fact that security is neither a given nor a historical norm.

In this sense, I treat trafficking as a symptom and not a cause of state vulnerability. The latter highlights important and persistent structural, institutional, and contractual deficiencies within the organization, management, and narration of the nation-state. Because trafficking is explained as a failure to protect oneself, as exploitation or deceit in the case of survivors, or the result of treachery, violence, and criminal intent in the case of perpetrators, it is understood as amenable to correction and arrested proliferation. This is followed by a process of careful documentation in the catalogs and annals of NGOs, academic publications, counseling centers, state institutions, intergovernmental organizations, newspapers, and international observer missions—a truly gargantuan remedying machine indeed. Thus, a trafficking survivor must make her/his way across international borders so s/he can be rescued, reported, deported, rehabilitated, and introduced back into legally sanctioned existence.

Trafficking necessarily both breeds insecurity and challenges the desire for security. This dichotomy legitimates a discourse that not only relegates trafficking to a privileged category of state intervention but also precludes certain kinds of engagements with survivors outside the boundaries of official directives. Because trafficking is understood as synonymous with insecurity, eliminating and combating it must necessarily create the conditions for security. Or, even better, the "war on trafficking" now raging initiates and justifies a war-like discourse aimed at restoring peace. Thus, stories of trafficking relayed through the language of war "distort . . . perceptions of risk. Reports of the most horrific cases generate excessive fears among women about the potential threat of violence from dangerous strangers,"[8] leading to an affirmation of the need to combat what is seen as an epidemic and vilify the perceived criminal elements that perpetuate it. However, what if eradicating and combating trafficking only produced further insecurity and defined individuals and communities away from belonging, empathy, and hospitality and closer to fear, uprootedness, and distress?

IN LIEU OF A LITERATURE REVIEW

Previous research determines and underwrites discourse, policy proposals, and the general ways in which trafficking and other "events" like it are understood, framed, and constructed so that a "securitizing speech act . . . [can] be accepted by a relevant audience"[9] without falling prey to misunderstandings, omissions, and lacunas. It is worth sharing the entire official definition of trafficking, taken from the Convention against Transnational Organized Crime out of the United Nations Office on Drugs and Crime (UNODC), also known as the Palermo Protocol.[10]

> (a) "Trafficking in persons" shall mean the recruitment, transportation, transfer, harbouring or receipt of persons, by means of the threat or use of force or other forms of coercion, of abduction, of fraud, of deception, of the abuse of power or of a position of vulnerability or of the giving or receiving of payments or benefits to achieve the consent of a person having control over another person, for the purpose of exploitation. Exploitation shall include, at a minimum, the exploitation of the prostitution of others or other forms of sexual exploitation, forced labour or services, slavery or practices similar to slavery, servitude or the removal of organs;
>
> (b) The consent of a victim of trafficking in persons to the intended exploitation set forth in subparagraph (a) of this article shall be irrelevant where any of the means set forth in subparagraph (a) have been used;
>
> (c) The recruitment, transportation, transfer, harbouring or receipt of a child for the purpose of exploitation shall be considered "trafficking in persons" even if this does not involve any of the means set forth in subparagraph (a) of this article;
>
> (d) "Child" shall mean any person under eighteen years of age.[11]

Almost identical to the Palermo Protocol is the EU definition, in which trafficking is "the recruitment, transportation, transfer, harbouring or receipt of persons, including exchange or transfer of control over that person, by means of the threat or use of force or other forms of coercion, of abduction, of fraud, of deception, of the abuse of power or of a position of vulnerability or of the giving or receiving of payments or benefits to achieve the consent of a person having control over another person, for the purpose of exploitation" in which "a position of vulnerability occurs when the person has no real or acceptable alternative but to submit to the abuse involved." Exploitation, in this respect, "shall include, as a minimum, the exploitation of the prostitution of others or other forms of sexual exploitation, forced labour or services, including begging, slavery or practices similar to slavery, servitude, or the exploitation of criminal activities or the removal of organs."[12]

Under U.S. law, sex trafficking is when "a commercial sex act is induced by force, fraud, or coercion, or when the person induced to perform such act has not attained 18 years of age" or "the recruitment, harboring, transportation, provision, or obtaining of a person for labor or services, through the use of force, fraud, or coercion for the purpose of subjection to involuntary servitude, peonage, debt bondage, or slavery."[13] All definitions share a focus on exploitation, vulnerability, deceit, and force. As long as one of those elements is present, especially when dealing with minors, and even in cases in which the survivor may have knowingly entered into debt, labor, and/or sexual servitude, trafficking is still considered to have occurred.

The most prevalent critiques of the protocol and its derivative treaties and agreements have to do with the complex wording of the definition,[14] its relative silence on domestic instances of trafficking, whether the latter should be seen as a crime when not in an international context,[15] its relative ambiguity on issues of consent,[16] and its equation of trafficking with "slavery."[17] While none of these critiques has been successful in invalidating or amending the protocol, if such an invalidation and nullification were even necessary or desirable, they have introduced persistent disagreements among trafficking activists. The ensuing counterproductive results can be related back to the negotiation processes that went into drafting the Palermo Protocol, in which the three prevalent views of trafficking as a migration issue, a criminal law issue, or a human rights issue inevitably clashed, producing a compromise that, to all sides, fell far short of producing an effective, applicable, and internationally useful document.[18]

More specifically, the struggle between the abolitionists looking to stop all instances of sexual labor, trafficking, and prostitution alike and the organizations part of the Human Rights Caucus committed to defining and treating trafficking as a human rights issue and away from a demonization of sexuality and sex labor compromised the final document in ways that additionally have impeded its effectiveness. During the drafting process, "The Human Rights Caucus consisted of an alliance of human rights, anti-trafficking and sex workers' rights organisations and activists, with a leading role for the International Human Rights Law Group (IHRLG) and the Global Alliance Against Trafficking in Women (GAATW)."[19] Proposing to view trafficking as a human rights issue took the focus away from the wrongdoing of the "victim" and toward a focus on denial of the right to work, travel, and be in control of one's body, among others. A human rights focus on the issue was also suggested by way of precluding the criminalization that followed the Palermo Protocol and that has offered largely punitive measures of addressing trafficking.

The second block, "led by the originally American based Coalition Against Trafficking in Persons (CATW) . . . and its partners, among which the European Women's Lobby (EWL) and the International Abolitionist Federation (IAF), regard the institution of prostitution itself as a violation of human rights, akin to slavery."[20] Thus, rather than focus on the deception and coercive nature of the labor performed, the abolitionists' moral claims define all prostitution as always forceful, evil, and exploitative, seeing sex workers as slaves who lack agency. Thus, prostitution and sex trafficking are inextricably linked to evil and to each other, making any attempt to legalize prostitution a contributing factor to a worse situation.[21] Out of this emerges a humanitarian emergency in which women and others need to be saved and in which the inherent purity of the female body is not only sacred but also antecedent to the purity and security of the nation-state. In opposition to this stance, the Human Rights Caucus insists on recognizing and respecting the right of men and women to engage in sexual activities for money without being villainized, shifting the focus away from sexual exploitation and toward recognition and protection of the rights to movement and employment.

In *Sex Slaves and Discourse Masters*, Jo Doezema provides a masterful analysis of the ways in which historical references to "white slavery" not only legitimized current abolitionist discourses but also initiated a humanitarian politics based on partial understandings of both slavery and trafficking and became responsible, as a result, for the perpetuation of narratives relating to trafficking and predicated on a savior logic.[22] Claudia Aradau offers a similarly ambitious engagement with trafficking in her *Rethinking Trafficking in Women*, in which she not only defines trafficking squarely as a security issue but also critiques humanitarian discourses on pity for producing a dichotomous universe in which trafficking interventions enter the register of humanitarian government as "a set of medical and confessional techniques" that administer life in such a way as to "render bodies and lives less risky."[23]

Additionally, Dominika Borg Jansson's *Modern Slavery: A Comparative Study of the Definition of Trafficking in Persons* questions the prevalent belief that trafficking must necessarily involve a passive, inactive, victimized person who is always, no matter the original intent either of the trafficked person or the trafficker, exploited, with undue attention being given to conditions in the country of origin, overlooking the conditions in the countries of destination. Her engagement with domestic laws in Sweden, Poland, and Russia shows inconsistencies and difficulties in unproblematically applying the Palermo Protocol to any national context, reducing the discourse to a one-sided engagement with trafficking primarily focused on the illegal crossing of borders and the criminality of recruiters and clients.[24] Particularly interesting is her discussion of the fact that while the Palermo Protocol sets up somewhat clear parameters of who and what fits under trafficking dependent on a determination of exploitation and lack of free choice, neither one can often be clearly proven, understood, or deduced, unduly limiting, under the rubric of necessity, "the understanding of trafficking to offences against *physically* coerced women."[25]

To make matters worse, the numbers related to trafficking further complicate things, as they contribute to the frequent conflation of trafficking with sex work, smuggling, illegal migration, and prostitution. According to the 2017 estimates by the International Labor Organization (ILO), forty million people were victims of modern slavery[26] in 2016, twenty-five million of which were exploited for their labor and fifteen million of which were forced into marriage. EU numbers, quoting the ILO, estimate modern slavery, including both labor exploitation and forced marriage, in 2016 to have affected around 24.9 million people. The EU numbers on trafficking are 2.45 million persons a year, with 1.2 million of them children (50 percent).[27] The ILO estimates that 71 percent of modern slavery victims are women and girls, with one in four being children (25 percent). What is more, "In the past five years, 89 million people experienced some form of modern slavery for periods of time ranging from a few days to the whole five years."[28] Additionally, the numbers Human Rights Watch (HRW) uses for their campaigns rely on ILO statistics, offering little room for critique of accuracy and scope. The EU uses the UNODC statistics, which has sexual exploitation at 79 percent of all cases in trafficking. The U.S. State Department further puts the number of people trafficked annually in the world at 800,000 to 900,000.[29] The explaining focus that a statistical engagement with trafficking offers falls short of doing much besides representing numerically a phenomenon acknowledged as hard to quantify and difficult to measure accurately.

The International Organization for Migration (IOM) 2006–2016 report released in 2017 indicates that while in 2006 84 percent of trafficking victims were female, that number dropped to 43 percent in 2016, meaning that male victims now make up more than half (57 percent) of all trafficking victims worldwide, with over three-quarters of the victims (77 percent) being trafficked over an international border. Their data put the average duration of trafficking at 2.2 years, with men remaining in trafficking on average one-third longer than women. Additionally, while in 2006 69 percent of identified female victims were subject to sexual exploitation, in 2016 that number had dropped to 28 percent, indicating an almost identical reversal in numbers when speaking of labor versus sexual exploitation.[30] Why the disparity? Maybe because not only is data on trafficking contradictory, unreliable, and inaccurate but they also boast sweeping and authoritative conclusions, paying very little attention to the intersection of consent, agency, and choice—issues that not only affect representations of trafficking but also the fate of so-called victims of trafficking.

The example of Peter, whom I met in Bulgaria and who was a "victim" of labor exploitation in Greece, Belgium, Austria, France, and then Slovenia, clearly points to the blurry lines between consent, exploitation, and physical abuse. Peter knew he was begging for someone else's financial benefit, he knew he was not going to keep all the money, he knew the whole time that what he was doing was illegal, and yet he engaged in street begging for years not because he left Bulgaria "choosing" to break foreign laws or because he "wanted" to do just and only this kind of work, but because he had no other options and because every other employment opportunity presented to him prior and after was either financially unattractive, equally exploitative, and/or simply dangerous. Begging was, to him, and because of a preexisting physical handicap, something he was good at and something that allowed him to send money home while leading a relatively carefree existence under the care of the man in charge, whom he called, lovingly, "boss." When I spoke to Peter, he was living in a hospice-like facility, having just been returned from

Slovenia, anxious to leave the country again to resume begging abroad first chance he got. Under the Palermo Protocol, Peter is a strange fit. He was not transported against his will, knew his job would be begging, and yet accepted the offer as one would accept any other legitimate employment opportunity. Even though he did not conceive of himself as having been exploited by his "boss," he qualifies as a labor trafficking survivor.

Additionally, the Palermo Protocol places undue focus on trafficking for sexual exploitation when, in fact, the latter constitutes merely 20 to 30 percent of all recorded trafficking cases per year worldwide. Despite a realization that the privileging of sex trafficking over other forms of trafficking is directly linked to the historical evolution of law, "the linkage between the two had led, in many instances, to overlooking the extent to which persons are trafficked for other forms of exploitation."[31] What is more, the risk factors involved with trafficking prevention and victim identification, defined as apolitical and largely associated with demographics and socioeconomic factors, work to delimit who counts as a trafficking victim and who does not, reducing the conversation on trafficking to one about numbers, in which rather than addressing trafficked survivors' economic needs, the latter are "constructed through the lens of psychological risk."[32] If neither risk nor security are given, how do they figure in discourses on vulnerability that plague potential trafficking victims?

THE EXPERT AS A KEEPER OF KNOWLEDGE

The risk that trafficking survivors incur is more than just "danger" or "insecurity" and should not carelessly, eagerly, or too easily be assigned to swaths of the world citizenry. The assumption that risk leads to trauma that necessitates ameliorative state mechanisms to be engaged with obfuscates the specificities of each trafficking survivor's story. Ulrich Beck is useful here when discussing "relations of definition" and their role in constructing a world "confronted with large-scale threats whose origin lies in the triumphs of modern society (more industry, new technologies), threats which, in view of the industrialized state promise of security, can nevertheless neither be adequately confirmed nor attributed, nor compensated, nor (preventatively) managed in accordance with prevailing legal, scientific, and political principles."[33] As a result, the global proliferation and definitions of risk are also the direct result of the workings of "relations of domination" predicated on "specific relations of definitional power"[34] that are, in spite of the different and varying nature of risks, static and designed to uphold a status quo that precludes addressing risk effectively. "Relations of domination," in this sense, are constructed as the continuation and maturation of "relations of definition" that are the product of "control over the 'means of definition'"[35] and inform as well as underwrite a hierarchy of knowledge privileging the "owners of the means of definition"[36] as the authorities deciding "on behalf of all which conflicting 'definitions of risk,' and which liability and compensation claims derived from them, are recognized and which are not."[37] Said "relations of definition" thus go on to produce various forms of domination expressed through the power of one group over others.

Ulrich Beck is useful again when discussing how the specific knowledge required for combating specific risks falls under the rubric of "organized irresponsibility" in which information is not shared, contradictory in its legality/illegality, and, on the whole, exposing "non-ascribability of the threatening effects of decisions" inside nation-states and the fragmented nature of "legal spaces between nation-states."[38] As a result of such rationalizations, "victims and perpetrators exist in a methodological duality . . . are divided into two categories: those who can be medicalized, controlled, and reformed and those who must be removed beyond the territory of a civilized society."[39] Trafficking is an example of this hierarchy of knowledge laying down the "superiority of the expert vis-à-vis the layperson,"[40] so much so, and following from literature on sexual abuse, that "social workers maintain that their expertise is indispensable"[41] in validating "the trauma model, as institutionalized in professional practice" and "essential to the process of verifying sexual abuse and providing evidence of it for use in the legal process."[42] Kristin Bumiller rightfully calls this "administrative justice,"[43] in which survivors' accounts are often substituted for the accounts of professionals about them, something I encountered frequently during my field work. Intent on protecting the victim, representing the victim, speaking for the victim, and even hiding the victim, social and governmental agencies I met with in Bulgaria defined my experiences in gaining access to survivors of trafficking within a strictly controlled network of captives and keepers in which the survivors were engaged in activities aimed at raising and solidifying their awareness of their self-worth, ability to hold a job, manage a budget, and carry out a variety of other everyday activities.

Raya, one of the survivors I met, spoke eagerly of her desire to move on, even after having had to give away her youngest child who was extorting her mother, threatening her with a knife, and mocking her less than perfect command of Bulgarian (Raya was Roma).[44] During our meeting, she spoke slowly and carefully, seeking help whenever she could not think of the proper Bulgarian word. We were accompanied by a social worker associated with one of the international NGOs in Bulgaria working with survivors of trafficking, who wanted to make sure I was not inappropriate, that my questions would not retraumatize Raya, and that if any "interpretation" was necessary, it would be provided for me. Raya was highly complementary of the work of the NGO in charge of her rehabilitation and thought their care exceeded what she deserved (a theme I encountered with other survivors I met as well). She expressed a desire to travel to the United States of America, not to Europe, where she thought things were bad. Though of Roma descent, she did not keep in touch with her family because she disapproved of their ways of life, which was something she was reclaiming for herself and something that she saw separated her from other Roma. She expressed a well-realized need to control her desires, her wishes, and her actions, which stood in stark contrast to her ex-husband, who was "addicted to sex and was also a homosexual."

By way of an example, she shared that she had recently bought herself a microwave with a grill, to the dismay of the social worker, who was supposed to approve and be informed of all bigger and nonnecessary purchases, all part of a budgetary plan they were executing together in the hopes that Raya would one day be able to take care of herself, hold a job, and adopt responsible spending habits. Raya was charming, smiled a lot, smoked a lot, and wanted to share her story—she said I was a good listener and predisposed her to talk, something she liked doing when given the opportunity to meet "good people." I was, over the course of our meeting, included in the list of those "good people" that Raya was praising at the NGO, so much so that the social worker who was accompanying her suggested I pursue social work as a career.

While social workers carry out a mandate of eradicating and mitigating trafficking trauma through recovery that is planned, their work is also undertaken somewhat independently of the larger underlying conditions that lead trafficking survivors to emigrate in the first place. The assumption is that survivors need to be retrained to hold a job, pay rent, and take care of themselves, as if their previous inability to do all these things was the main contributor to ending up in trafficking. For example, Maria was given the opportunity to get training to become a hairdresser so she could get back on her feet. After completing the training and being unable to find work as a hairdresser, she was forced to work in a small stand (one the size of a phone booth) selling lottery tickets and cigarettes. The training was offered for free so she took it, as did Elena, though for neither did it provide a long-term solution to unemployment, nor did it offer a viable alternative to seeking more gainful employment abroad. The training itself provided an empty promise, turning the responsibility for finding employment back to survivors, absolving the nation-state of responsibility for supporting a meaningful rehabilitation process.

It would be erroneous to claim that social workers and counselors are willing accomplices in some grand scheme of victimizing and diminishing the confidence of survivors, when the former themselves, as experts when it comes to trafficking, are victims to already defined expectations about the survivors' "problems, needs, and patterns of recovery."[45] Social workers and counselors are, after all, dependent on the state for their funding and for legitimizing their organizations' efficacy in the wider community of trafficking professionals. Their own attitudes toward the trafficking survivors are an important consequence of their understanding of widespread attitudes regarding the subject of their professional expertise.

A world in which a "rational male" can make choices as an opportunistic migrant overconstructs women/girls as "victims" and overlooks men, boys, and other victims who do not fit the trafficking victim profile.[46] The risk that trafficking survivors bring to the state derives, especially for women, not only from their willing or unwilling participation in criminal activities but also from their desire to cross an international border. In that sense, no movement guarantees a risk-free existence, when in fact the trafficking survivors I interviewed saw movement and traveling for work their best chance at reducing the risks that delimited their lives. Yet trafficking survivors remain collectively risky to the nation-state also because their "capture" by bad guys showcases a failure on the part of the powers that be to ensure security in which "the effort to 'outlaw' social injury powerfully legitimizes law and the state as appropriate protectors against injury casting injured individuals as needing such protection by such protectors."[47] Thus, the urge to reify trafficking experiences from a position of moral certitude leads to a reality in which they are entirely constructed but also interpreted, explained, and made legible without end. Renewing its focus on bad actors, "the dominant group reinforces its self-perception as savior, even while providing little real assistance to most trafficking victims or those who might be at risk of being trafficked."[48] This programming of individual identities mirrors programmed state identities underwriting all understanding of risk and vulnerability from the point of view of national priorities. Daring to propose a framework away from legal and political readings of violence, Étienne Balibar asks, "Must we not seriously doubt whether the state, *on its own*, is really an agent of civility?"[49] and further adds, "The nation is, ultimately, an *impossible* entity, which can never entirely achieve its ideal, and it is as such—that is to say, as a problem—that it is *real*."[50]

As a result, the international gaze focused on identifying, rescuing, and rectifying the situation of trafficking survivors and other "vulnerable populations" has fallen especially on women, constructed as susceptible to all sorts of indignities and dangers lurking in a big, scary world. These dangers, in effect, become the prevue of law enforcement, the legal system, and professionals in the counseling and human services sectors of governments, in which survivors and those assisting them exist in an "objective reality" of sorts in which "the rule of law is advanced as a necessary component for just societies."[51] The role of the social worker becomes one of guarding knowledge and bodies, where survivors of trafficking in human beings are carefully sheltered and shielded from any outside intrusions.

The kind and welcoming ways in which I was greeted in conducting all my field research were examples of "relations of domination" that resulted from social workers and other responsible parties' "relations of definition" regarding my place in their world. In effect, the logic of institutional help does more than just "victimize" survivors of trafficking: "We must, then, suppose that the role of institutions is precisely to reduce—without suppressing—the multiplicity, complexity and conflictuality of identifications and senses of belonging, if need be by applying a preventive violence or a 'symbolic' and material—corporeal—organized counter-violence."[52] So within the world of experts itself there exist clear boundaries of inquiry that informed both my ability to access the survivors themselves and my ability to inscribe what I did learn in the register of political testimony.

TRAUMA, TRAFFICKING, AND THE LAW IN BULGARIA

While it is true that trafficking could befall anyone, we are not all, despite terrorist, economic, environmental, racial, sexual, gender, existential, and other threats, equally insecure. We do all carry, however, the potential of insecurity. How is insecurity from trafficking different from other kinds of insecurities? Emotions, in the context of this chapter, are privileged as "*reflexive of bodily memory*"[53] and thus central to one's understanding of and engagement with survivors of trafficking. In the end, the goal of learning is the creation of "places of memory" in which people "give shape to felt absence, fears, and desires that haunt contemporary society."[54] Put otherwise, in addition to holding an emotional charge for the survivor, memories of trauma, as in the case of trafficking, also contain a deep political charge.

For example, in the context of Bulgarian antitrafficking law, after a survivor has been identified, the first meeting with authorities, though intended to be simply an informational one, is often an interrogation. Survivors are not informed of their rights and prosecutors are able to gain valuable information to make their case in court. In many cases, the survivor does not know whether s/he is a witness, a plaintiff, or a defendant in a case. Often, the survivor is seen as both a witness and a victim whose rights need to be recovered.[55] What is more, while there is a small group of lawyers trained and authorized to work with survivors of trafficking, the quality of their work is not checked, and the actual legal aid they offer survivors of trafficking is often not readily available. In addition, most trafficking survivors don't ask for a lawyer when they meet with police because 1) they don't know they can ask for one; 2) they don't realize their vulnerability while being interviewed; and 3) they are afraid to ask for one, seeing themselves as guilty and not as "victims."[56] Last but not least, in the Bulgarian context survivors of trafficking are not obligated to cooperate with the prosecution, but if they choose not to, rights and services are denied them, including free stay in shelters for survivors of trafficking run by the state and free legal and social consultations, in a quid pro quo kind of arrangement. Survivors become trapped in a system that needs their cooperation and also their obedience. This is another example of Wendy Brown's thesis that states validate their claims to legitimate control with recourse to actions that deprive others of their rights.[57] Ironically, one becomes a survivor only when inscribed in the legal system.

What is more, most judges in the Bulgarian context defend the thesis that in order for the truthfulness and the facts of a case to be established, a survivor's sexual past needs to be interrogated so clear understanding regarding his/her "willful" or "forced" participation in any trafficking for the purposes of sexual exploitation may be determined.[58] Yet according to the Palermo Protocol, for trafficking to have occurred, consent by the survivor need not be established and is not antithetical to coercion, exploitation, or violation of rights. In other words, having left the country knowing one will be working in the sex industry does not exonerate traffickers of blame for lying, misrepresentation, and exploitation in the form of garnishing wages, unacceptable working conditions, and confiscation of documents.

The inability of trafficking survivors to find employment in their home countries, the strict and getting stricter border regimes across Europe, the difficulty of finding legitimate and gainful paid employment abroad, and the revictimization that survivors of trafficking undergo post-trafficking all help explain trafficking survivors' choice not to cooperate with the justice system. Often, in doing so, they risk being labeled as amoral or "perverse" in conjunction with Bulgarian laws defining prostitution as "deviant" behavior, even though it is not explicitly criminalized according to the country's criminal code. The conclusion by one Bulgarian judge relating to the criminalization of prostitution and ensuing trafficking can be summarized as follows: lacking specific and explicit wording criminalizing prostitution in the country, and lacking humane conditions of labor under a different occupation, the responsibility for seeking employment in the sex industry cannot immediately fall on the shoulders of the person engaged in the "amoral" action.[59] Trafficking, in this sense, is as much a labor as a sex issue and must be discussed in light of questions of work, wages, and rights. In other words,

> provisional or not, the situation is that millions of disposable human beings are at the same time excluded from labour—that is, *economic activity*—and kept within the boundaries of the *market*, since the market is an absolute; it has no *external limits*. The Market is the World. When it excludes you, you cannot leave it, search for another America, settle there and start again.[60]

Drawing on Beck's comments on the mutually constitutive relationship between "relations of definition" and the ensuing "relations of domination"[61]—which is the resulting operationalization of power in the political, personal, and social spheres—Todres's analysis of alterity[62] adds to our understanding of the Palermo Protocol by insisting that the effectiveness of international law is hampered by its inability to account for and address the discursively and legally endorsed discrimination underwriting said "relations of domination." The goal here is not to eliminate power but to understand its functioning.

> A society without power relations can only be an abstraction. Which, be it said in passing, makes all the more politically necessary the analysis of power relations in a given society, their historical formation, the sources of their strength or fragility, the conditions which are necessary to transform some or to abolish others. . . . Instead I would say that the analysis, elaboration, and bringing into question of power relations and the "agonism" between power relations and the intransitivity of freedom is a permanent political task inherent in all social existence.[63]

Thus, the law reflects "societal expression of what is valued—and thus international law that expresses a lower priority on assisting marginalized persons reaffirms the otherness—shaped perspectives and values held by citizens."[64] As a result, "by focusing on prosecuting bad actors, the dominant group reinforces its self-perception as savior, even while providing little real assistance to most trafficking victims or those who might be at risk of being trafficked."[65] The group of interest here, the group whose control of the "relations of definition" with respect to trafficking determines prevalent "relations of domination," is the sovereign nation-state, and the Bulgarian government in particular.

BULGARIA AND THE EUROPEAN UNION ON TRAFFICKING

As already alluded to, a similar ordering of reality and ideality, law and criminality, Self and Other also happens with respect to Bulgaria's relationship to EU directives on trafficking, the latter assuming the role of evaluators and curators of Bulgaria's ascension and positive contributions to the EU. On the one hand, experts want to empower, as has been shown in the preceding pages. On the other hand, EU directives often dissuade from that goal by privileging national and international priorities rather than the priorities of the survivors. It is unclear, however, looking at the Council of Europe document on trafficking monitoring, otherwise known as the Group of Experts on Action against Trafficking (GRETA) report,[66] what the goals of the EU are in this case. Bulgaria is cast by the European community as a project that is evaluated periodically. Recommendations are given concerning its dealing with, reporting on, identifying, and managing of trafficking as such. For example, the GRETA report published on January 28, 2017, states that even though the number of Bulgarian nationals rescued from sexual exploitation

(77 percent) is still the highest of any other category, begging and labor exploitation (12 percent in 2015) are gaining visibility as well. The report notes that about 45 percent of people who were eventually trafficked across international borders from Bulgaria were first internally trafficked. The 2016 Report by the Bulgarian National Commission for Combating Trafficking in Human Beings (NCCTHB)[67] is aligned with the GRETA report, providing easy cross-referencing. Before I proceed with a brief analysis of both reports and their findings, allow me to mention that survivors of trafficking in human beings are, in both reports, referred to as "victims" repeatedly.

In fact, other than specific gender, race, and age identifiers used to distinguish among different categories of people affected by trafficking in human beings and for the purposes of categorization, statistical differentiations, and specific recommendations, the term "victims" is the predominant way these individuals are addressed and assessed, creating "a rhetorical trope" to depict trafficking as "an object of humanitarian action, law enforcement intervention, and human rights policy."[68] Alongside Jo Doezema's research elucidating the connections between "white slavery" and "human trafficking" in terms of the constructed nature of the discourse of saviors of victimized women,[69] the use of the term "victim" in the 2016 official report by the Council of Europe on trafficking in Bulgaria is not only outdated but also counterproductive to the recommendations for empowerment, assistance, and combating trafficking to which they claim to be committed. While I recognize that a discussion of terminology won't help any survivor of trafficking in human beings get back on his/her feet, it draws attention to the depoliticized ways speaking and addressing survivors of trafficking victimize the latter further.

In the Bulgarian context of alleviating suffering, the 2017 (covering activities in 2016) and the 2018 (covering activities in 2017) NCCTHB reports do a painstaking job of listing all the various ministries and organizations engaged in the fight against trafficking in human beings, each of them participating in trainings and educational seminars on a regular basis aimed at gaining tools and knowledge to be more effective in combating trafficking. Meetings, international initiatives, UN summits, and more trainings are listed in what seems to read as an endless trail of bureaucratic undertakings. Not a single time is a survivor of trafficking referred to by name, even a pseudonym, and survivors are routinely called victims. The aspirational and self-congratulatory tone of the report reads like the bureaucratic document that it is and nothing more. Time and again, minority status in rural areas and economic insecurity are mentioned as the main driving factors behind trafficking from Bulgaria, but both economic insecurity and one's ethnic, minority, and social status are somehow left as givens—neither one is problematized, nor does the report concern itself with unpacking what "economic insecurity" in rural environments really means or whether anyone is responsible for alleviating said insecurity from within the government. The findings relating to statistics, demographics, and vulnerable groups are perhaps even correct but do nothing to report on "trafficking" as a phenomenon that affects the individual human body. Yet the reader is comforted reading, in bold, that Bulgaria has the highest capacity for helping survivors of trafficking. What this capacity means and who Bulgaria is compared to in this praise, however, remain unknowns. Prosecutors are educated, lawyers are trained, judges go to meetings and seminars, and multidisciplinary working groups are formed and spread their wings on the pages of the report, but little specifics are provided regarding results of their work, accomplishments, and, if one may be so bold as to suggest, critiques.

The GRETA report recommendations are echoed by the NCCTHB assurances that things are being done and that lots of people are involved in trying to resolve the trafficking issue in a manner satisfactory to the Council of Europe. The language in both relies heavily on the work and contributions of "experts" and "authorities" who come together at "high-level meetings" and "training sessions" in which "evidence-based research," "research plans," and "needs assessments," combined with "preventative work," "action plans," "raising awareness campaigns," and "vigorous measures," all contribute to promoting, among other things, "gender equality, combat gender-based violence and stereotypes, and support specific policies for the empowerment of women as a means of combating the root causes of THB."[70]

With regard to specific assistance measures available to survivors of trafficking in human beings (THB in the report), the report lists state-run as well as NGO shelters in the country, various initiatives undertaken by volunteers, the regional Commissions for Combating Trafficking in Human Beings, the Ministry of Justice, the National Agency for the Child, courts, the Ministry of Internal Affairs, the Ministry of Foreign Affairs, the High Cassation Commission, the Ministry of Labor and Social Policy, the National Employment Agency, the Central Labor Regulation Commission, the Ministry of Health, and the Agency for Social Welfare and Support—and the list goes on. Then the report goes over all the nongovernmental and supra-national organizations involved in addressing the gathering of information and the rehabilitation and return home of survivors of trafficking. In some detail is also engaged the work of regional commissions in nine Bulgarian towns, which carry out regional educational campaigns on the topic as well as provide immediate assistance and information to survivors and their families.

When Bulgaria and other Eastern European countries are evaluated by the Council of Europe, which produces the GRETA report, a critical yet uncritical discourse is produced and disseminated, necessitating disciplining actions in response to how one country is faring against other countries. Yet Bulgaria is not invited to comment on West European countries' handling of survivors of trafficking once they are apprehended in the West; neither is Bulgaria allowed to make suggestions as to how a more effective, open, and democratic border regime on the part of Western European countries could, in fact, alleviate the proliferation of trafficking and provide employment opportunities for women and men, minimizing the risk of exploitation, discrimination, oppression, and violence. Thus, the country of citizenship and origin becomes the main target of criticism while the countries of destination remain the arbiters and main referees within a disciplining regime. It would be a mistake to suggest that the Council of Europe simply intends to patronize the Bulgarian and other governments by monitoring their work related to trafficking in human beings. But is the goal of GRETA to make sure the Bulgarian government is doing what it "should" be doing even though human trafficking is still proliferating around the world and in Bulgaria? Is it to shame the Bulgarian (and other) government for not putting enough efforts into combating trafficking? Is it to come up with a benchmark standard for engaging with the causes, proliferation, and combating of trafficking for both origin and destination countries?

Some of the report's specific recommendations include the establishment of an independent rapporteur who will be able to evaluate the progress made toward meeting the goals of the current National Trafficking Action Plan. Additionally, the GRETA report encourages more training, human investment, and educational opportunities for all parties involved in working with survivors of trafficking. The Bulgarian authorities are encouraged to continue "sensitizing relevant officials, in particular labour inspectors, prosecutors and judges, about THB" as well as "strengthening the monitoring of recruitment and temporary work agencies," complimented by "social, economic and other measures for groups vulnerable to THB, including outreach work in Roma communities" as well as "efforts . . . to promote gender equality, combat gender-based violence and stereotypes, and support specific policies for the empowerment of women as a result of combating the root causes of THB."[71] Additionally, the report strongly advises the Bulgarian authorities to pay attention to illegal foreigners in the country and to their vulnerability to being trafficked while in immigration detention centers. A conversation I had with a representative from the Agency for Refugees produced a strong assurance that there never have been nor were there at the time cases of trafficking among their clients that I, at the time and still today, find very hard to believe.[72]

A strong recommendation at the end of the report critiques the Bulgarian government's treatment of children survivors of trafficking in human beings by identifying as wrong "the institutionalization of children" and encouraging "the development of a fully-fledged system of family and community-based services, to reform the juvenile justice system with no further delay and in particular repeal the obsolete Law on Measures against Delinquency of Minors and Juveniles, and to move away from the current repressive approach towards a protective one."[73] It is obvious by looking at some of the more explicit recommendations relating to the treatment and care of children survivors of trafficking in human beings that their risk of being retrafficked is directly related to their care after rescue. The report refers to the need to assess whether it is safe for children to be returned to their parents, the need to ensure safe reintegration, the need to provide further training to stakeholders, the need to establish communication "with relevant destination countries and EU institutions with a view of adopting effective cross-border mechanisms of identifying children at risk to be re-trafficked," as well as "taking steps to address the problems of unaccompanied minors disappearing."[74]

But what are the tools through which democratically elected officials in a country that is largely struggling to provide for its own people, a country that aspires to be European, a country that is considered the poorest of all EU member states, and a country that struggles to accept, treat as equal, and assimilate its Roma minority supposed to use to implement GRETA recommendations, even if they were willing to do so? In the grand scheme of things, trafficking is a minor problem to most Bulgarians and, while recognizing its existence, they would be hard-pressed to pay too much attention to it. Unless, of course, it affected someone they knew. In such a climate, it is not surprising to read the pronouncement of a judge who refused to grant material damages to a survivor with the argument that "sex work brings 'immoral income' and the victim had no legal ground to claim compensation from the trafficker."[75] Considering that "there were five cases in which eight victims were granted compensation for nonmaterial damages in the period 2008–2012" and that "there is no information about any civil claim submitted by a trafficked person before a civil court," ensuring that "all victims of human trafficking, regardless of nationality and residential status, are eligible for State compensation," not only does the ability of the state to address this issue seem like a tall order of a task but a highly unrealistic one too.[76] Even though one of the 2013 amendments to the Bulgarian Criminal Code included the insertion of Article 16a, reading "an act shall not be considered culpably committed if performed by a person who is a victim of human trafficking and who was forced to perform the act in direct relation to being such a victim,"[77] the interpretation of said provision did not impede the judge quoted above from dismissing a claim to material compensation on the basis of "immoral behavior" while trafficked.

Like law, "presented as objective reality" in which "the rule of law is advanced as a necessary component for just societies,"[78] trafficking encompasses a system of legal rules historically endowed with legitimacy that "promote equality for some and permit, or even promote, enslavement or exploitation of others."[79] In this sense, understanding where vulnerability comes from, what it means, and how it is perpetuated starts with examining the nation-state as the organizing structure of law and society, and less with writing reports on the myriad meetings conducted and planned on the question at hand. In other words, and returning to Doezema's use of myth to describe the construction of trafficking tied to a certain ideology, the belief that there are certain social arrangements that are beneficial for the community necessitates, as a response to the need to maintain said social arrangements, certain discursive and real policies to be promoted. The Bulgarian anxiety with not being European enough likewise exposes an unfulfilled desire to belong, creating a reality in which certain iterations of lawfulness and acceptable behavior are dominant not because of facts but because of prioritized ideological commitments. It is useful to refer to Doezema's words in full here:

> The ironic power of the myth of trafficking is that in attempting to lift the mask through exposure of the "facts," it only settles firmer. The face appears to be the true face but as Laclau demonstrates, there is no "true face," only the impossible but necessary space enacted by ideology and sutured by myth. The myth of trafficking persuades us that we are on the right course: we investigate the facts, denounce the distortions and triumphantly claim that the myth of trafficking is no more—exposed to the harsh light of truth. Yet, our every step has been at the behest of myth. We imagine ourselves free of trafficking at the moment we are most captured by it.[80]

And thus follow the legitimation, naturalization, and legalization of trafficking as a political discourse/reality, all the while operationalizing and interpreting individual stories away from their diverse and multiple meanings and toward a unity of fact, experience, and ideology. On the flip side, "facts never *speak* for themselves but are always *made* to speak by a network of discursive devices,"[81] often ushering in metaphoric violence.

After Wendy Brown's insightful analysis, "freedom neither overcomes nor eludes power; rather, it requires for its sustenance that we take the full measure of power's range and appearances—the powers that situate, constrain, and produce subjects as well as the will to power entailed in practicing freedom."[82] In this sense, the security promised to survivors of trafficking is not aimed at providing freedom but is self-referential and largely disciplinary, defined by "the powers against which it arrays itself as well as the power it must claim to enact itself."[83] The promise of a secure subject thus becomes inscribed in discourses on (in)security, in which a specific "we" as a referent object of security is constructed and reconstructed. Maria Stein offers an interesting analysis of this dynamic in her engagement with (in)security narratives of Mayan women, in which the enactment of local security narratives against an encroaching world is done both on the part of the state and on the part of the women, in which the "we" that is secured necessarily is constructed as "fixed in space through the geographic, cognitive and political borders of the state/nation/political community."[84]

The institutionalization of care, the operationalization of crimes, and the appropriation of stories are all instances of "victim and villain" narratives, in which oversimplifying the reality of trafficking contributes to an oversimplification of reality as such. The power of this myth thus lies in the totality of its explanation in which rough edges of experiences are rounded off and smoothed over. Looked at structurally, a cultural myth is a discourse, "a set of narrative formulas that acquire through specifiable historical action a significant ideological charge."[85] As a result, certain discursive, social, and political practices are endorsed as legitimate, desirable, and morally superior, supporting specific beliefs about the world while rejecting others. "Political myth in this conception of ideology is a weapon of the powerful, a falsehood through which the dominant power can legitimize and consolidate its position."[86]

In this chapter, risk and vulnerability have been engaged with outside the realm of brooding, deviant, dangerous, and insatiable predators inhabiting a universe in which the "inevitable" proliferation of cross-border crimes in a globalized, post-Communist Europe is on the rise. Mariella Pandolfi offers an interesting engagement with the state of exception in the Balkans, especially with relation to the fall of Yugoslavia and the ensuing violence, arguing that "managing life in the Balkans in the name of human rights" can be seen not simply as "a response to an exceptional historical accident" but rather as an example of one of many "paradigmatic features of our times."[87] The social construction of states of vulnerability, risk, and danger is perfectly exemplified in the case of the trafficking survivor thus: "The 'disposable human being' is indeed a social phenomenon, but it tends to look, at least in some cases, like a 'natural' phenomenon, or a phenomenon of violence in which the boundaries between what is human and what is natural, or what is post-human and what is post-natural, tend to become blurred: what I would be tempted to call an *ultra-objective* form of violence, or *cruelty with a Medusa face*."[88]

NOTES

1. Vicky Squire, *Post/Humanitarian Border Politics Between Mexico and the US* (London: Palgrave McMillan, 2015).
2. Claudia Aradau, *Rethinking Trafficking in Women* (London: Palgrave McMillan, 2008).
3. Ibid., 91.
4. Tugba Basaran, "Security, Law, Borders: Spaces of Exclusion," *International Political Sociology* 2, no. 4 (2008): 340.
5. And this other could, potentially, refer to men and women who have been trafficked, the fear of the proliferation of trafficking in the indefinite future and what it might do to the state, the fear of the kinds of behaviors that trafficking exposes that must then be rectified, and the fear of being unable to contain the subject within acceptable boundaries of behavior.
6. Basaran, "Security, Law, Borders: Spaces of Exclusion," 340, 341.
7. Wendy Brown, *Walled States: Waning Sovereignty* (London: Zone Books, 2010).

8. Kristin Bumiller, *In an Abusive State* (Durham, NC: Duke University Press, 2008), 18.
9. Aradau, *Rethinking Trafficking in Women*, 68.
10. The Palermo Protocol's full name is "Protocol to Prevent, Suppress and Punish Trafficking in Persons Especially Women and Children, supplementing the United Nations Convention against Transnational Organized Crime."
11. Palermo Protocol, http://www.ohchr.org/EN/ProfessionalInterest/Pages/ProtocolTraffickingInPersons.aspx (accessed November 4, 2017).
12. https://ec.europa.eu/anti-trafficking/citizens-corner/trafficking-explained_en (accessed November 5, 2017).
13. https://www.humanrightsfirst.org/resource/human-trafficking-numbers (accessed November 5, 2017).
14. Dominika Borg Jansson, *Modern Slavery: A Comparative Study of the Definition of Trafficking in Persons* (Leiden: Brill Nijhoff, 2014).
15. Jansson, *Modern Slavery*; Jonathan Todres, "Law, Otherness, and Human Trafficking," *Santa Clara Law Review* 49, no. 3 (2009).
16. Jansson, *Modern Slavery*.
17. Jo Doezema, "Loose Women or Lost Women? The Re-Emergence of the Myth of White Slavery in Contemporary Discourses on Trafficking in Women," *Gender Issues* (Winter 2000): 23–50; Jo Doezema, "Ouch! Western Feminists' 'Wounded Attachment' to the 'Third World Prostitute,'" *Feminist Review* 67 (2001): 16–38; Jo Doezema, *Sex Slaves and Discourse Masters: The Construction of Trafficking* (London: Zed Books, 2010).
18. Doezema, "Loose Women or Lost Women?"; Doezema, "Ouch!"; *Sex Slaves and Discourse Masters*; Jansson, *Modern Slavery*.
19. Melissa Ditmore and Marjan Wijers, "The Negotiations on the UN Protocol on Trafficking in Persons," *Nemesis* 4 (2003): 80.
20. Ibid., 81.
21. Ronald Weitzer, "The Social Construction of Sex Trafficking: Ideology and Institutionalization of a Moral Crusade," *Politics and Society* 35, no. 3 (2007): 447–75.
22. Doezema, *Sex Slaves and Discourse Masters*.
23. Aradau, *Rethinking Trafficking in Women*, 36.
24. Jansson, *Modern Slavery*, 92.
25. Ibid., 89.
26. In the ILO vernacular, modern slavery is understood as a combination of forced labor *and* forced marriage.
27. European Commission, "Together Against Trafficking in Human Beings," https://ec.europa.eu/anti-trafficking/citizens-corner/trafficking-explained_en (accessed November 5, 2017).
28. International Organization for Migration, "Forced Labour, Modern Slavery and Human Traffickig," http://www.ilo.org/global/topics/forced-labour/lang--en/index.htm (accessed November 5, 2017).
29. U.S. Department of State, "Trafficking in Persons Report 2017," https://www.state.gov/j/tip/rls/tiprpt/2017/index.htm (accessed November 5, 2017).
30. International Organization for Migration, "Global Trafficking Trends in Focus," https://www.iom.int/news/un-migration-agency-marks-11th-eu-anti-trafficking-day-new-data-analysis-human-trafficking (accessed on November 5, 2017).
31. Todres, "Law, Otherness, and Human Trafficking," 10.
32. Aradau, *Rethinking Trafficking in Women*, 100.
33. Ulrich Beck, *World at Risk* (London: Polity, 2009), 30.
34. Ibid., 30.
35. Ibid., 33.
36. Ibid.
37. Ibid.
38. Ibid., 31.
39. Bumiller, *In an Abusive State*, 13–14.
40. Beck, *World at Risk*, 33.
41. Bumiller, *In an Abusive State*, 71.

42. Ibid., 32.
43. Ibid., 63.
44. Author interview with Raya on May 10, 2017, in Sofia.
45. Bumiller, *In an Abusive State*, 68.
46. Todres, "Law, Otherness, and Human Trafficking," 9.
47. Wendy Brown, *States of Injury* (Princeton, NJ: Princeton University Press, 1995), 27.
48. Todres, "Law, Otherness, and Human Trafficking," 12.
49. Étienne Balibar, *Politics and the Other Scene* (London: Verso, 2002), 33; emphasis in original.
50. Ibid., 65; emphasis in original.
51. Todres, "Law, Otherness, and Human Trafficking," 9.
52. Balibar, *Politics and the Other Scene*, 29.
53. Giorgio Hadi Curti, "From a Wall of Bodies to a Body of Walls: Politics of Affect/Politics of Memory/Politics of War," *Emotion, Space and Society* 1 (2008): 107.
54. Karen Till, *The New Berlin: Memory, Politics, Place* (Minneapolis: University of Minnesota Press, 2005), 9.
55. Natasha Dobreva, *Protection of the Rights of Victims of Trafficking in Human Beings* (Sofia, Bulgaria: Foundation Association Animus, 2013), translation by author; *Deliana Markova*, interview by author, Sofia, Bulgaria, March 2, 2017.
56. Ibid.
57. Brown, *Walled States.*
58. Dobreva, *Protection of the Rights of Victims of Trafficking in Human Beings*, 67.
59. Ibid., 84.
60. Balibar, *Politics and the Other Scene*, 142.
61. Ulrich Beck and Elizabeth Beck-Gernsheim, *Individualization: Institutionalized Individualism and Its Social and Political Consequences* (London: Sage, 2001).
62. Todres, "Law, Otherness, and Human Trafficking."
63. Hubert Dreufys and Paul Rabinow, eds., *Beyond Structuralism and Hermeneutics* (Chicago: University of Chicago Press, 1983), 222–23.
64. Todres, "Law, Otherness, and Human Trafficking," 11.
65. Ibid., 12.
66. Report Concerning the Implementation of the Council of Europe Convention on Action against Trafficking in Human Beings by Bulgaria, GRETA (2015), 32 (accessed January 30, 2017).
67. National Commission for Combating Trafficking in Human Beings (NCCTHB) Yearly Report 2016 (accessed February 20, 2017), translated by the author.
68. Gretchen Soderlund, "The Rhetoric of Revelation: Sex Trafficking and the Journalistic Exposé," *Humanity* 2, no. 2 (2011): 193.
69. Doezema, "Loose Women or Lost Women?"
70. Report Concerning the Implementation of the Council of Europe Convention on Action against Trafficking in Human Beings by Bulgaria, 23.
71. National Commission for Combating Trafficking in Human Beings (NCCTHB) Yearly Report 2016, 20–23.
72. Bulgarian Agency for Refugees representative, author interview, Sofia, Bulgaria, April 4, 2017.
73. National Commission for Combating Trafficking in Human Beings (NCCTHB) Yearly Report 2016, 29–30.
74. Ibid.
75. Report Concerning the Implementation of the Council of Europe Convention on Action against Trafficking in Human Beings by Bulgaria, 32, 35.
76. Report Concerning the Implementation of the Council of Europe Convention on Action against Trafficking in Human Beings by Bulgaria, 32, 35–37.
77. National Commission for Combating Trafficking in Human Beings (NCCTHB) Yearly Report 2016, 29–30.
78. Todres, "Law, Otherness, and Human Trafficking," 9.
79. Ibid.

80. Doezema, "Loose Women or Lost Women?" 45.
81. Slavoj Žižek, *Mapping Ideology* (London: Verso, 1994), 11.
82. Brown, *States of Injury*, 25.
83. Ibid.
84. Maria Stein, "'We' the Subject: The Power and Failure of (In)Security," *Security Dialogue* 37, no. 2 (2006): 193.
85. Doezema, *Sex Slaves and Discourse Masters*, 31–32.
86. Ibid., 36.
87. Mariella Pandolfi, "From Paradox to Paradigm: The Permanent State of Emergency in the Balkans," in *Contemporary States of Emergency: The Politics of Military and Humanitarian Interventions*, ed. Didier Fassin and Mariella Pandolfi (New York: Zone Books, 2010), 154.
88. Balibar, *Politics and the Other Scene*, 143; emphasis in original.

Chapter Four

Recovery, Pathology, and Poetic Witness

The witness-based nature of recounting trafficking experiences offers the perfect avenue for exploring alternative ways of engaging with, interpreting, and translating that witness for others. The approach, commitments, and goals of the singular witness lens accommodate a witness/survivor whose subjectivity can be restored away from the realm of "othered subjectivity." What is more, and what this chapter argues, is that, through poetry, "the process of witnessing is not reduced to the effects of trauma. So, too, subjectivity is not reduced to the effects of trauma"[1] or to one's ability to make this trauma known because the latter can erase subjectivity just as readily as it is charged with granting it. Therefore, how we choose to welcome, listen to, comment on, and engage with both the event and the telling of it means committing to impartiality and nondiscrimination in our methodological approaches. How can a novel way of engaging and translating survivors' testimonies help inform an alternative understanding of both relations of domination and relations of definition?

In a field in which visible violence on and over the body is largely privileged over violence inherent in the nation-state, an engagement with trafficking through poetry provides an alternative venue for engaging with other invisible traumas, in which "human as representation" gives way to "human as experience." The savior function that the state assumes must be questioned with respect to the actual ability of the state to provide security for individual citizen(s). Security is thus not synonymous with "freedom," as controlling and curbing the freedom to move around and travel is central to the state's goal of securing the body and is easily translated into a desire to reduce and, if possible, eliminate risk and vulnerability altogether. Freedom and security are rather posited as oppositional and necessarily at odds with each other.

Risk, by extension, becomes the mitigating factor, the thing that erroneously and misleadingly seems to unite the two, though in a reverse way: rather than tying security to freedom, security is predicated on erasing ambiguity and anxiety and thus freedom as well.

Establishing the conditions for creating and sustaining a political plurality that allows for feelings of empathy and compassion to entangle us with the "other" necessarily starts with a commitment to a "radically embodied, contextual, relational ontology"[2] transferred through stories. Storytelling uncovers and honors the excess of the singular, irreducible experience of trafficking as such. This "relational ontology of plurality and uniqueness"[3] opens the possibility for recovering the conditions for a political community both defined and challenged by the singularity (nongeneralizability) of worldly experiences. "No one may be liberated or elevated to a position of equality—let us say, may *be emancipated*—by an external, unilateral decision, or by a higher grace. Only reciprocally, by mutual recognition, can this be achieved."[4]

> A baby cries in a hammock
> peeling paint from dry walls.
> "It was terrible
> I don't want to talk about it."
>
> "The NGO gave me a bed
> for the baby and I
> but because I don't have a room
> it's in the barn.
>
> Sometimes we dust it
> and the other day mother put a blanket on it
> to dry herbs. Flat and high above ground."
>
> The houses around, crumbling, grin in the April sun.
> The Gypsy King shows me his kingdom,
> his barefoot vassals kicking a cloth ball
> at dawn of a new age—the age of Jesus.
>
> Beyond a turn, three palace-like homes stand out of place,
> yellow and pink walls behind tall fences
> guard a lone donkey in the front yard
> while a million potholes laugh at a Jeep and an SUV.
>
> "I could have had a house like that," says the King,
> but I would never have forgiven myself for sending my only daughter
> to sell her body
> for money . . .
> I have morals, even if I am Gypsy."

The face of anti-sex trafficking campaigns
turned victim of sex trafficking
sits on a bed in the barn,
cupping pregnant belly,
pulling straw out of her hair:

a queen robbed of her throne—
bunches of herbs for a dress.[5]

This representation of a meeting I had with a survivor of trafficking[6] shows "what bodies *do* rather than . . . what they *are*."[7] Bodies are undone and redone in the context of their relationships with other bodies and, in the process, reposition our understanding not only of the political bearings they inaugurate but also of the emotional engagement with the "hard stuff" that makes up their histories and their traumas. Here, "the role of emotion as an adhesive of experience cannot be separated from that of memory: it is not either memory or emotion that adhere together blocs of experiences in (re-)creations of identity, but both together working through a reflexive embodied symbiosis."[8] In this sense, I follow Marita Sturken's affirmation that the body "is a technology of memory through which memories are shared, produced, and given meaning"[9] and where memories, especially from traumatic experiences, are dependent on "continual acts of perception and recollection"[10] evoking a process and a reality of intersectionality, exchange, and a "gnawing at the present."[11] It is this gnawing at the present that I wish to honor hereafter.

PETER

For Peter, a late middle-aged man I met on a rainy day in Sofia, the lack of employment opportunities in Bulgaria played a big part in his becoming a "victim" of labor trafficking and then subject to the Bulgarian government's saving grace. When I met Peter, he was staying at an old people home until the organizations that were in charge of his rehabilitation and reintegration could figure out what to do about his situation. He was surrounded by bedridden individuals, all dying or close to dying, in a place that smelled of pee and sweat and had a "guard" at the door. The main nurse's incessant smoking was almost a welcome respite from the stench, and the word "sanatorium" would not have been an exaggeration by way of naming the place. Peter was alone in his room when I walked in; he had been expecting me, though he did not know who I was or why I was there. Someone from one of the NGOs caring for him had called to let him know a lady would be coming to talk to him, but his lack of comprehension about why I was there and what my project was about made me uneasy. Our conversation was not only awkward but also painfully unequal. What struck me the most about Peter was his fear, his sense of being lost and in a place where he did not belong, though he never settled on a place where he would have belonged or where he would have liked to be instead. He smoked a lot, was a very small man in his late fifties, though he looked at least sixty-five, and was dark skinned but not a Gypsy; he insisted on that several times. His story was not unusual, yet it is worth sharing in full.

He had gone to Greece to work as a shepherd, a job arranged for him by his brother, who was already working and living in Greece. That had not worked out as his employer had not paid or treated him well and refused to send money to his family. Unfortunately for Peter, he also lost his Bulgarian ID card that allowed him to travel around the EU without a passport. As a result, he quit the job and went to the Bulgarian consulate in Athens either to try and get a temporary card so he could resume work or to figure out a way to return to Bulgaria on a short-term travel permit. While passing the weekend waiting for the Bulgarian consulate to reopen on Monday, he found himself at the Athens train station, where he knew he could sleep for free and not have to spend a lot of money. Once there, he resorted to begging so he could make money for food and cigarettes. That is when he met a Bulgarian couple and their young child, his employers for the next few years.

His story recounts how they offered him shelter at their nice apartment, how they bought him clothes, food, alcohol, and cigarettes, and gave him his own room, an allowance, and a job: begging at busy street intersections in Athens. Peter said he was good at what he did and after a few weeks became even better. He bragged that he was not afraid to walk between the cars and did not get too tired from the job. It was the first time during our conversation that I saw him liven up, his face almost breaking into a smile, like the one children get when they manage to trick their parents in a clever way. He took another drag from his cigarette and proceeded. He praised his "boss," the affectionate term he used for his trafficker, for his humane attitude, nice house, and the fact that every time Peter asked for some money to be sent to his wife, the man did that for him at no cost. He allowed Peter breaks, gave him what Peter considered a fair share of the money, and treated him like a member of the family. It was surprising to hear someone genuinely so indebted to his "trafficker," signaling a potential disconnect between dominant perceptions and definitions of trafficking and the reality of it for some of its survivors, a reality of meaningful relationships of mutual economic and emotional benefit. Peter had a self-content way when he spoke of begging, proud, if you wish, of his abilities and his knack for tricking people for money.

Peter's story continues: after a while, another man joined him at his boss's house, a Bulgarian he did not like very much but with whom he learned to live and whom I shall call Georgi. Peter admitted to drinking and to his boss buying them alcohol when they asked for it at no charge. This was a good, if unexpected, development, according to Peter, because he was cared and provided for in multiple ways. Eventually, they all went to Brussels for three months, then on a short stint to Vienna, and finally to France, where they joined the boss's older brother, who had his own begging ring in Paris. Once in Paris, Peter said things turned sour. The older brother was cruel, did not want to give the men the same share of the money they had been receiving before, and entered into disputes with his younger brother and others in the camp as a result of feeling like his leadership position was under threat. Peter said his time in France was miserable and that he asked to be taken to Bulgaria, wanting to end his employment. It was also during that time that his Bulgarian ID card disappeared again. As a result of his precarious legal situation, the younger brother agreed to drive Peter to Bulgaria as he had some other business to attend to there. While on the way, he dropped Peter off in Slovenia, along the side of the highway and near a busy truck rest stop, while he was attending to some business nearby. Peter explained that this was supposed to be only temporary and that the brother was scheduled to return for him, had the police not picked him up first. His encounter with the police seemed to him simply an unfortunate circumstance, not something for which his "boss" could be held responsible.

During his three-day stay at the truck stop, Peter resumed begging, having been left only with a rucksack and little money for food. He said begging along the highway was good because he was able to make a lot of money and, what is more, keep all of it for himself. This was highly exciting for him, and he said how proud he was to buy his beer(s) with money he had earned himself and that he did not have to share with anyone. He said, "Some days I would only work a few hours and then could stop, without anyone telling me otherwise. Then I would buy my drink, sit down in the shade and have it, because I could and wanted to."[12] I can still vividly hear the drawl in his speech, as if he was singing while talking. Yet it was really his many missing teeth that were making his speech musical. When the police apprehended him, he was questioned and gave the names of both brothers to the authorities, mostly because he was scared and because he did not want to get in trouble with the law. A "victim" of trafficking, the law was on his side.

Peter's story is unique because he had a lot of regrets about giving away his "boss," who he felt was a person who genuinely cared for him, had a wife and a small child, and was now in jail because of Peter's confessions. He was afraid people were coming to get him, that his security could never be guaranteed again, and that the nongovernmental organization in charge of him was not helping him get in touch with his family so that he could ask to be taken out of the hospice. He seemed disoriented, spoke of lacking medicine to treat his epilepsy, and did not know how long he would stay in the facility (information that I ironically had but could not disclose), what would happen to him after that, and how he would make it out of Bulgaria again alive. He spent his days looking outside the window at cars' license plate numbers, sure that the brothers' hit men were coming for him. Smoking obsessively, he called me "Mrs." over and over again and, considering I was almost half his age, his reverence and insistence on a respectful and deferential tone when addressing me signaled low self-confidence and self-worth, as well as inflated respect for figures of authority. I must have seemed like quite the accomplished lady to him, what with my recorder, notebook, expensive raincoat (we talked in the courtyard in the rain, the only private place in the whole hospice), and ability to provide things he needed, like the cigarette packs I bought him after our meeting. I could not help but think of Peter as a prisoner, trading one sentence in the form of begging for another as an unemployed and scared, not sacred, trafficking survivor.

Peter represents an example of what Zygmunt Bauman calls "wasted life,"[13] someone who is not needed, is expendable, redundant, and altogether unfit to survive in the modern capitalist, liberal-democratic, and, as Étienne Balibar adds, "post-institutional" world.[14] Zygmunt Bauman starts *Wasted Lives* with a story about Ivan Klima's dinner with the president of the Ford Company. To Klima's question regarding disposal of used-up and broken-down cars, "he replied that this was not a problem. Anything that was manufactured could vanish without trace, it was merely a technical problem. And he smiled at the thought of a totally empty, cleansed world."[15] If Peter and the others I spoke with exemplify similarly broken-down, used-up lives and if their experience of trafficking relegates them to a position of abjection, how does a politics of "*order-building*" explain away this "human waste" while, at the same time, addressing Bauman's profound realization of "*an acute crisis of the human waste disposal industry*."[16] These "wasted lives," Bauman argues, are not accidental consequences of modernity but rather "an inevitable outcome of modernization."[17] In other words, this "surplus population" is made up of people who are "superfluous, unnecessary, unneeded and unwanted, and their reactions, off the mark or absent, render the censure a self-fulfilling prophesy."[18] Peter was one of those intruders, his life "hovering on the brink of criminality" as he was "feeding parasitically on the social body,"[19] raising "alarms about deteriorating security which magnify already plentiful 'security fears' while simultaneously shifting public concerns and the outlets for individual anxiety away from the economic and social roots of trouble and toward concerns for personal (bodily) safety."[20]

Yet his body was a site of power and, in important ways, exposed the reparatory and social processes he was subject to as a necessary part of legitimizing the state.[21] Techniques of domination unleashed by the state victimized him further and over these, he had very little control, as had not been the case with his begging. He was victimized in the hospice because of his lack of knowledge, his vulnerability, a lack of social community with which he could interact, and his financially precarious position. In effect, sovereign security depends on the excessive numbers of such individuals: an excess that can politely be tolerated so long as there are enough of them, as Richard Rorty says, "'who get their hands dirty cleaning out toilets' while being paid ten times less than we 'who sit behind desks and punch keyboards.'"[22]

Moving beyond a straightforward equating of trafficking survivors with yet another population that can be made to fit within Giorgio Agamben's *homo sacer* framework, yet another population reduced to "bare life," yet another population defined through a lack and a strictly physical existence stripped of political agency in the name of order and security, this chapter argues that the trafficking survivor occupies myriad legal, illegal, and extra-legal sites and uncovers a singularity that exceeds the exception. "*The sovereign sphere*," for Agamben, "*is the sphere in which it is permitted to kill without committing homicide and without celebrating a sacrifice, and sacred life—that is, life that may be killed but not sacrificed—is the life that has been captured in this sphere.*"[23] Inmates on death row would be one example of individuals relegated to this sphere. Enemy combatants, illegal migrants, suspected terrorists, enemies of the state, and convicts of various sorts would all easily inhabit this realm too. Should the list keep getting longer?

In the framework of this chapter and the book as a whole, reference to Agamben's work is not obligatory but nonetheless useful for establishing a framework for discussing trafficking in the context of risk and borders. One's past trafficking experience(s) could very well be positioned within a discourse of "the state of exception" as discussed by Agamben. On the face of it, trafficking survivors are *homines sacri*: people who can be killed but not sacrificed because of their extra-judicial status, existing as it is either outside the law or in a space that is not subject to the law.[24] Upon further investigation, I wonder whether the category of *homo sacer* can indeed by located here, and, if so, how it adds to our understandings not only of the theoretical and ontologically significant positioning of trafficking survivors but of their actual relationship with nation-states charged with protecting and rehabilitating them.

Similarly, the process of bearing witness is not simply contained within the juridico-political terms of the *state of exception* and must, in order that it remain possible, address the question of recognition beyond juridical modes of engagement. Much like Judith Butler's argument quoted in Kelly Oliver's *Witnessing Beyond Recognition* that "no speech is permissible without some other speech becoming impermissible,"[25] for Agamben the ability to bear witness becomes circumscribed within an economy of subordination of linguistic to juridical fact that engages the subject only from within a realm defined by the workings of sovereign power. While, as Butler recognizes, the ability to engage the world in language can be coercive,[26] the choices the subject makes vis-à-vis language cannot be understood simply as directed by economies of power external to the self. If that were the case, then violence would be inescapable.

Testimony, as Shoshana Felman has suggested, is *performative* and not just *cognitive*, as far as it "strives to produce, and to enable, *change*."[27] It is a "relation of events," "bits and pieces of a memory," which figures as a "discursive *practice*" that does not, however, make up "a completed statement, a totalizable account of those events."[28] The human being in a *state of exception* as a nonwo(man) figures mostly in relationship to what cannot be said, defining the process of bearing witness as a lack rather than as a possibility every time unfolding anew, on its own terms. Subordination, in the sense in which it destroys an autonomous subject, also precludes the process of witnessing. To bear witness is to commit to an address and, at the same time, to be undone by the impossibility of a final address. In the same way in which the *state of exception* posits as opposites law and emergency, the sovereign and the subject, language and silence only so that it would have to choose one over the other, so does an engagement with a traumatic past that has been normalized foreclose the ability to bring memory of more than traumatic events into language.

IT'S A WAR

I argue that trafficking has been viewed as a pathology, a disease to be cured, examined, and studied both by clients and by well-wishers, commodifying even the most well-intentioned initiatives and ushering in a never-ending cycle of compliance measures, statistical studies, and a politics of savior humanitarianism that perpetuates insecurity and a narrative of a desirable and healthy nation. A politics informed by risk underwrites and legitimizes trafficking as pathology, in which what is constructed as pathological is dealt with and addressed via various remedying technologies dedicated to the eradication of insecurity, vulnerability, and contingency. In the process, discussions on the nature of power that, in turn, lead to the possibility of problematizing said power are all but erased in favor of "making live" the body. This pathological nature of politics itself, the diseasing that comes when power perverts legitimacy, is neither engaged with nor examined and becomes a justification for the fight against all danger, difference, and contingency. Trafficking survivors are thus, unsurprisingly, not only "others" but also othering to us normal citizens, what with their dirty past, their stories of exploitation, their neediness, and their *trafficked bodies*.

If security locks public discourse in a Pandora's box, it is imperative to recognize that "to decide what views will count as reasonable within the public domain . . . is to decide what will and will not count as the public sphere of debate."[29] Wendy Brown is informative again when she challenges us "to develop new spaces in which to decide politically, collectively, what is good, just and right, derived not from identity-based notions of 'who I am' but from a new ethics of 'what I want for us.'"[30] To be trafficked, thus, may not necessarily mean to be deprived of security, as both trafficking and security extend beyond concrete examples of vulnerability and thus force a recognition of an inherent, existential state of not being secure. Yet this assumed dispossession and insecurity inform deeper debates about concepts of self/other, deserving/undeserving, and sacrifaceable/unmournable lives.

In this sense, one way of understanding the singular, unique, and nongeneralizable experience lived through as a trafficking survivor is perhaps through a story of post-trafficking life. Taking lead from Étienne Balibar again, it is important to underscore the singular nature of experiences of trafficking and thus the necessarily generalized missing the mark when treating and engaging trafficking survivors through an institutionalized approach. "In so far as the concepts we have discussed here concern politics, they can be articulated only on individual pathways (or, more precisely, at the meeting-point of individual pathways). Such pathways, like truth, are necessarily singular; hence no model exists for them."[31] This becomes even more imperative as Bulgaria, in light of its trafficking problem, exists on the radar of the EU as a case generating macro- and "micro-emergencies." This, in turn, demands and requires organized and constant interventions corresponding "to a specific tool kit of operative practices specifically tailored to the particular segment of transition that needs to be addressed."[32] In this sense, the disorder that trafficking and its victims signal becomes the subject of international and local action plans operationalized in the work of national, transnational, and international governmental and nongovernmental organizations. The management, salvation, and governance of "victims" by the latter fundamentally work to produce and reproduce victimhood in eternity, whether they be victims of "war . . . corruption, underdevelopment, inadequate self-organization, failed markets, and the like."[33] As a result, the figure of the enemy, whether an "inner enemy" or an "outside foe," "creates favorable conditions for heightened social anxiety to be displaced and transferred onto minorities and marginalities,"[34] in which anxieties, a waning sovereignty, and fear all combine to produce the enemy as "an ever-changing political strategy."[35]

In a similar move, the Balkans themselves become parsed out and demarcated depending on their membership in the EU or their aspirations of belonging to it. They become a space separating desirable and undesirable members of the EU, in which unemployment, changing governments, inflation, rising prices, and falling salaries/pensions define the norm, not the exception, of being a Balkan citizen. In such an environment, not only have insecurity and risk been heightened, but they have also been inscribed on the backs of national self-loathing and deprecation, in which the need for better governance is supplemented by the need and call to blame someone for the lack of one. Survivors of trafficking in human beings expose an underlying inability of the Bulgarian government to protect, provide for, and keep safe its citizens. At the same time, what I found were many accounts of self-blame or blaming the survivors for their inability to make it like "the rest of us do" and not get trafficked or abused, transforming self-loathing into "the articulation of hate as a communal expression of humiliation and rage."[36] The kinds of interventions alluded to here do not constitute military or humanitarian ones but are carried out by observers, report writers, and NGO meetings aimed at setting the tone, inscribing for the Bulgarian side a model of behavior, both preemptive and post-crisis, that would somehow help show 1) that the country is making an effort at accounting for and tracking trafficking and its "victims" and 2) that the country is able to rehabilitate the latter successfully under the practice of "good governance." Even though "communication between different multilateral sectors may be inefficient and even contradictory,"[37] the trafficking machine operates despite lack of clear consensus between law enforcement, the judiciary, NGOs, and governmental organizations. Against this backdrop, Elena's story helps explain trafficking as the realm of the singular story.

Elena was quiet
and kindly accepted a Fanta.

Next door, in the park, was a small zoo:
mostly domestic animals, but also a small tiger—
dirt floor in cages stripped of anything
alive, blooming, colorful. Budgetary issues.

"Spaghetti," she said, "I like to cook spaghetti
and go for walks or short day trips."

"Tell me about your children.
Are the mountains as high as they say they are?
Don't you get scared driving
in the winter? I would be."

Hear the peacocks . . .

"Ivan and I would like
a baby one day
but not like the one I lost,
before . . .
he doesn't know
any of it."

A lonely ostrich stares through holes in the fence . . .

"Ivan is an orphan, unwanted—
how many truths can a couple live with?
I take care of him, he takes care of me
and since we have each other,
he doesn't need to know."

The tiger roars,
a monkey cries out in response:

Laws of the jungle.

Elena, alongside Paul Kottman in his introduction to Adriana Cavarero's *For More Than One Voice*, makes clear that "every utterance is moreover an action, which at once manifests one's embodied uniqueness to others in the context of a material, ontological relation here and now and which also initiates, or alters, a relation whose sense exceeds this sheer manifestation of one's corporeal uniqueness."[38] Elena's goal in life was a peaceful existence in the context of a family of her own.[39] When I spoke to her in the spring of 2017, she was living in a government apartment for orphaned children and had little opportunity to change her circumstances without incurring significant financial burden. She shared that she enjoyed her work of tending to the parks, alleys, and other green spaces in her home town and thought that being outside and with other women forced her to get out of her head, engage in small talk, and have a purpose. She seemed genuinely glad to be working outside, though to most other people her occupation would seem exhausting, underpaid, and hard work. When I asked what her favorite thing to do in her spare time was, she said it was cooking, taking walks, and reading. She shared that she had been reading two hundred books a year but recently had not had time to go to the library because of work.

To the question regarding how she was coping with sadness, she said she did not think it useful to focus on traumatic things in the past and that what she hopes for most of all is what she dreams about for her future. Elena was trained as a beautician as part of a post-trafficking job training program, and said that while she would like to work in that field, she would also be happy to do agricultural work (planting potatoes was one example she gave) or clean houses. Elena's experiences with her own family were not exactly positive: her mother had died when she was eleven, leaving her in the care of her father and two brothers, the former often unhappy with her because she was not always making the choices he wanted her to. She said she had felt wronged when her father left their family home to her brother who had since been extorting her for money. As a result, she had retained a lawyer and gotten her brother to sign an agreement not to extort her any more or threaten her boyfriend with beatings. Despite these challenges, Elena exuded a sense of calm, and before I left she said, "I wish I could tell other girls that money isn't everything, to be careful and not to trust everyone." While we were walking around the small animal park close to where we met, she also said this: "I think a lot of the animals are lonely. That makes me sad. If I ever decided to have a pet, I need to be sure I have time to take care of it; it is a huge responsibility." Positioned amid various state apparatuses, Elena was grateful for the supportive role the state had played after her "rescue" and seemed to appreciate the care of the social workers, as well as other survivors, she had come in touch with. Yet she was critical of a system that did not provide for its people and thus erected material Gods for its citizens, who were all too happy to flee and work in the sex industry, sometimes leaving small children behind. When asked whether she thought what had happened to her was exceptional or unusual, she laughed and said, "Nothing is exceptional. That's just how the world works."

Trafficking survivors interrogate the very legitimacy of law, sovereignty, and the register of international rights and obligations not because they can be and are killed/hurt without legal recourse to justice but because their experiences problematize the relationship between vulnerability, the nation-state, and the human body. In other words, trafficking victims' political identities constitute a series of negotiations and renegotiations of belonging and legality that cannot simply be reduced to one's experience as a trafficked person but must be understood as constituted by the myriad experiences, memories, and realities of being human. To that add the myriad technologies, organizations, and apparatuses involved in and responsible for rehabilitation, reintegration, and representation of trafficking survivors. This process of coming in and out of fixed political identities, which ironically makes them the exemplary global citizens, is constituted along several axes: first, when they make a decision to migrate as a member of a nation-state, crossing international borders marks them as bearers of political rights inscribed on travel documents; then, if and when trafficked, they cease to enjoy the protection of a sovereign center and enter a realm of illegality in which they are reduced to bodies for rent/sale, a predicament that makes "leaving" captivity difficult; again, when rescued, they undergo a period of temporarily belonging to several nation-state jurisdictions, bearing witness to their exploitation before foreign as well as "home" law enforcement officials; and when repatriated/deported, their position as a "victim" is again underwritten with reference to suffered coercion, forced captivity, and violence. I argue that the numerous identities imprinted on trafficking survivors indicate a broken politics, and a broken politics breeds closure and insecurity.[40] When asked to describe themselves as trafficking survivors, here is what I was told:

I am a woman
young
unemployed
uneducated
poor
Roma.

I am a mother
middle aged
unemployed
educated
poor.

I am a wife
educated
rural
indebted
beaten.

I am a man
middle aged
disabled
not Roma
employed.

I am a husband
father
provider.

I am a body
hard-working
good-looking
dream-seeking
used
useful
usable
on the move.

The parallel backstories this poem alludes to both offer a window into the experiences of trafficking survivors and complicate said experiences. While some people are more vulnerable to being recruited for trafficking than others, a "victim profile" as such does not exist, and thus focusing on trafficking education alone is sure to fail. What most trafficking survivors I spoke with identified as a main push factor was their indebtedness and poverty that, paired with lack of opportunities for gainful employment in Bulgaria, made them into perfect candidates for work abroad—sometimes ending up in trafficking, other times not. Beyond policy decisions, directives, and budgetary constraints unfold the lives of individuals for whom "insecurity" of one sort or another has, for a long time, been the driving force of decision making. Burdened with the responsibility of taking care of families, they resort to their own ingenuity, hard work, and perseverance as an avenue of ensuring more stable lives. Nobody had left because they thought waiting tables in the Netherlands would offer them the experience of a lifetime. Rather, as "vagabonds"[41] abandoned in some fashion by the nation-state, they each evaluate risk in terms of its relative bearing on their current insecurity. Ironically, this move also disposes of the nation-state whose role as provider is no longer sought after, and turns the gaze to instances of "everyday security" that the human engages in and which are sampled throughout this project.

In all these instances, trafficking survivors are managed, regulated, and objectified in one way or another by traffickers, academics, counselors, and law enforcement officials alike, so much so that they don't need to speak for us to know how they are feeling, having become subsumed in a larger economy of "ideal" trafficking utterances. Here is an example of a counter-metaphor:

> Coffee tastes best on a rainy day
> when cigarette smoke envelops the cup
> still hot, but cooling, from waiting.
> the woman across the table smiles.
>
> She tries for her story
> ending up, always,
> talking about paintings in living rooms,
> the shower faucet leaking,
> friends having dinner on balconies lit like Christmas trees,
> the sound of tramways at night,
> how to make bread, how to find the library
> and whether to take that job
> across the border.

In this sense, the subject of trafficking is no longer just a mythical subject "constructed through sexual [and other] vulnerability"[42] but becomes a narrating, human and political subject, problematizing straightforward assumptions about how trafficking happens and what it means.

HOMO TRAFFICUS: TRAFFICKING AS STATE MAKING

> Can we find another meaning, and another possibility, for the decentering of the first-person narrative within the global framework?
> —Judith Butler, *Precarious Life*, 7

It is difficult not to write about trafficking, and at the same time it is difficult to write about it. In the same way that all humans may experience, at some point or another in their lives, an existential condition of loss, homelessness, and angst, so do most human beings *never* experience being trafficked—and that is something that one either lives through or does not. It is not a condition that can be inferred or adopted. What can one say about something that is, at the same time, so personal and so public, so difficult to control/combat and so easy to proliferate illegally, so seemingly visible and yet so misunderstood and hidden, so widely researched and yet so misrepresented and opaque? The tendency to assign victimhood to all trafficking survivors itself necessitates a closer look at the phenomenon and a more careful examination both of the practices that drive it and of the ways we speak and write about it. The trafficking survivors I spoke with all wanted their stories to be told so that, in the words of Irina, "others would not go through what I went through." Each thanked me for my interest in their stories, though they did not all know what trafficking was, nor did they all feel exploited and abused, but of that in a bit.

If economies of power affect who says what, when, and how, then all witnessing/speaking is a political activity and carries the mark of a politicization that does not take as its reference point the juridical that contains within itself Agamben's exception. In this sense and contra Agamben, the nature of the political that is of interest here is contained within everyday acts of speaking and address that are not predicated either on the law or on the sovereign inclusion that brings the former into the realm of the exception. "As a performative speech act, testimony in effect addresses what in history is *action* that exceeds any substantialized significance, and what in happenings is *impact* that dynamically explodes any conceptual reifications and any constative delimitations."[43] The setting up of documented narratives about the past affects the ways in which histories are written; the passing of judgment on the past affects the ways in which ethical and legal categories are established; the decision about who can say what, when, and how affects the ways in which assignments of identification are constructed; the establishing of guilt, innocence, and culpability affect the ways and modalities through the medium of which the law sets up everyday practices of interaction between the sovereign, the citizen, and the law. Concerning the relationship between calculability and bearing witness, for what information can the survivor be responsible? In the register of the body, the process of witnessing and its relationship to truth telling is hardly ever only an *exception*, an excess, or a lacuna (lack).

In the Agambenian register, the identification of a person or a population as *homo sacer* relegates said subject to a position of powerlessness in which the moment of exploitation turns into a lifetime of normalized oppression defined importantly by lack of access to rights and a legal status. Yet could it be that positioning someone as a *homo sacer* preclude political action and resistance? While the category of *homo sacer* has great explanatory value, does it provide a way of escaping the very apparatus responsible for marginalization, of identity seeking away from its grips? Vicky Squire points to a similar potential in Agamben's thought to contribute "to the abjectification, abnormalisation and depoliticization of those who are subject to the exclusionary dimensions of contemporary bordering practices."[44] The implications that follow from being identified as *homo sacer*, which terrorist suspects, refugees, the homeless, and criminals have been equated with time and again, helps tease out the myriad ways in which the formal power states exercise over the bodies and lives of individuals can also inspire resistance. It is this invisible, at first, but possible and potentially transformative resistance to which I believe trafficking survivors point.

Famously, the Jewish population (alongside Gypsies, homosexuals, and political dissenters) during the Holocaust embodies perfectly the systematic dehumanization that contributed to the legal and state-endorsed extermination of millions of individuals defined as "inferior," "unworthy," "enemy," and "other." In Agamben's analysis, the sacred human can be killed, and is sometimes killed, but cannot be sacrificed and, as a result, remains unmourned[45] and central for establishing and justifying a "state of exception" in which the temporary suspension of the law is instituted as the new norm outside/beyond law. Might the potential for resistance itself be contained precisely in these extra-juridical spaces to which the "other" has been relegated? If so, what are the consequences for engaging in such acts of resistance beyond mere speaking out? Could the exclusion prompt an examination of the potential to resist the exclusion in which *homines sacri* aren't simply static, vulnerable, and, like the mainstream conception of the trafficking victim, terribly wronged, humiliated, and used? The "politically irrelevant"[46] life that the sovereign gets to decide on, in Agamben's world, seems to remain so in perpetua. Might the trafficking survivor offer an exception to the exception? In other words, *homo sacer* depoliticizes the subject even more in a kind of perverse irony. If what is understood and defined as dangerous is now all that exceeds the parameters of law-abiding and state-sanctioned behavior, what of the possibility to critically engage with the particularities and singularities of different life stories and experiences?

So how does trafficking fit within Agamben's work? Dolores Chew neatly connects Agamben's *homo sacer* to trafficking by equating trafficking survivors to migrants who "have a biological life but no political significance."[47] To recognize and admit that a group or an individual lacks political agency, which is something that can be said for many, many minority or quasi-minority groups in many, many societies and states around the world, is itself only partially politically significant and certainly not empowering in any meaningful way. In a world in which "the state is indirectly and inescapably the source of vulnerability,"[48] it is this state (of) vulnerability that is largely and primarily responsible for the proliferation of insecurities befalling individual citizens and that, ironically, calls up the need for protection by the institutionalized apparatus of the state. The political and cultural prominence this sovereign move has earned the discourse on trafficking is not to be underestimated, as it ushers in a kind of "carceral politics and a securitized apparatus to be . . . the preferred political remedies."[49]

The body of the trafficking survivor thus comes to be owned by the trafficking discourse. In this sense, definitions and identities are created outside discussions of the norms and discourses producing said definitions, taking for granted that a survivor of trafficking in human beings while also a parent, spouse, or worker must be addressed primarily as a survivor of trafficking in human beings. As Kristin Bumiller shares, "It has become nearly impossible to understand the causes and consequences of being a victim of violence in terms which do not fit squarely within the purview of medicine or criminal justice."[50] As national security evolves from reworking an amalgamation of insecurities, the body of the trafficking survivor becomes one of those insecurity-perpetuating obstacles. In the same ways in which the sexual violence crisis has been anchored around discussions of sexual deviance, a trafficking crisis is making its way into the public consciousness in which what befalls young innocent women must be identified, punished, and criminalized. In this way, "the production of a crisis then serves as a justification for state intervention to regulate sexual practices."[51] The naming of trafficking as an epidemic only serves to reinforce the idea that foreign travel is dangerous and that sending countries are involved in perpetuating a kind of pathology of lawlessness throughout Europe and beyond. One of the unintended, or perhaps intended, consequences of this move is restricting the ability of women to travel in search of work. While providing for their families is seen as an important task for men, female migration is decried and women's independence is seen as a threat to the integrity of the family unit and the unity of the nation, further scrutinizing their freedom of movement[52] and thus, ironically, underwriting oppression.

The pathologization of trafficking thus not only gives nation-states the ability to lay claim to all modes of restitution but also justifies and necessitates greater scrutiny when applying for "help" as a trafficking survivor, leading to a reality in which women "understand their contact with the system as part of their process of victimization."[53] As a result, not only is women's freedom of movement limited by having to take precautions regarding their social, personal, and economic lives because of where danger comes from—mafia-like tough guys, sex tourists, the Albanians, the border, employers, other countries, the police[54]—but their interactions with the system further preclude opportunities for "meaningful choice and autonomy"[55] as survivors of trafficking often exist in hierarchical relations with professionals or others within the legal system. It is a master plan that is predicated on "saving bodies"[56] in which a "real" victim joins forces with an "ideal" victim, in which "real" stories underwrite *the* "ideal story" of a collective trafficking experience.

The trafficking victim represented as "vulnerable" also challenges the concept of vulnerability itself, and in informative ways. Survivors of trafficking draw attention to the ways in which vulnerability and risk are conceptualized and carefully orchestrated by state apparatuses and institutions, in which sovereign fragility is interposed on and supersedes individual vulnerability. Trafficking, in this sense, enters the register of conditions bestowed upon the human by states and inscribed against the backdrop of measures to ensure the security of the whole state governmentality mechanism, for the benefit of citizens on the inside against outsiders. Can vulnerability and being at risk be seen not as states of being per se but as conditions of possibility in which trafficking survivors, understood and engaged with by people making policy decisions about their rehabilitation, can exceed and outlive their vulnerability through stories of post-trafficking life? The pervasive and overwhelming difficulty and challenge that all trafficking survivors I spoke with experienced were the daily struggles to find work, an affordable and safe place to live, and the ability to "shed" their trafficking victim identities. The insecurity derived from their status as victims of trafficking was perpetuated by a regime of rehabilitation that, if not entirely, was largely predicated on the creation, study, and management of pathological humans.

In other words, "the trauma model, as institutionalized in professional practice, became essential to the process of verifying sexual abuse and providing evidence of it for use in the legal process."[57] In the context of this research, the trained professionals working on behalf of the state or within the institutional boundaries of states engage in an economy of silencing, sometimes unbeknownst to them, sometimes as willing accomplices, expressed in "the complex ways in which those who claim authority and expertise about the issue authorize what can be said about sexual violence"[58] or trafficking, by extension. Perhaps unavoidable, this complicity is unsettling and deserves a careful interrogation, if only to rescue trafficking survivors from the fate of data.

For example, when I spoke with Peter he insisted on having worked hard his whole life, having always been poor but an honest man. What he did (begging) in Greece, Belgium, Vienna, Paris, and then Slovenia was, for him, simply an extension of his work history and something he was good at because he was able to make a lot of money for his "boss" and his own family. When I spoke with Irina,[59] she indicated that begging in Sweden had been amusing to her; she was able to observe people, talk to them, and even befriend some of them. She did not think her job was excruciating, though it wasn't particularly rewarding or mentally stimulating either; she was not treated badly, though she did admit that looking back at it, she realizes she was exploited, denied her labor rights, and treated in a manner less fair than if she had not been in a foreign country and utterly dependent on her trafficker. For example, her boss asked her and her co-beggars to turn over any items given them by Swedes, especially drinks or things of value. She also said she was made to pay back any money she used to buy herself a soda while at work, which she also considered exploitative, and that going to the bathroom too often was also frowned upon. Yet Irina, because she was older, had had more success in claiming rights for herself. She shared that this was not true of the men on the team and of younger women, especially the one who "fell in love" with the boss.

This positioning of the self as a close accomplice, confidant(e), and even love interest vis-à-vis one's exploiter was, according to Irina, a way for her female friend to cope with the situation. Irina said that the group she worked with, who all lived in an old barn outfitted to house people, were a close-knit community, all of them Bulgarians who helped each other when necessary. They did not beg at the same stores and only saw each other at night, yet spent holidays together and cleaned and kept up their "place" daily. When their "boss" was apprehended by the police, they were not arrested with him but left in the "barn" to continue working, if desired, because begging is legal in Sweden. Because their electricity and hot water were shut off and winter was coming, they got together, figured out what bills had to be paid, which amounted to pretty much all the electric and other bills for their stay there, and split up their earnings in order to cover the costs. She said being able to beg and keep all the money gave them a sense of freedom, even if it was only short lived. In the end, however, they figured they could not make it on their own, contacted the police, and asked to be returned to Bulgaria.

To critique Peter and Irina's work ethic would be inappropriate, yet it was their work ethic that allowed the traffickers to benefit from them. They were disciplined, obedient, and honest, doing what they would have done in Bulgaria or elsewhere if given the opportunity to hold a job that made them money. Their employment abroad was seen as "an opportunity" and not necessarily as a criminal act, and they understood that in order to be able to make money, they would have to give in to others' demands. They both called their employers "boss"—an indication of the clear hierarchical demarcations of their relationship. Neither one, especially Irina, saw their stint in begging as something that would define the rest of their lives, though she did find it peculiar that she had been a "beggar"—herself a poet, mother, and property owner in Bulgaria. Had they had the opportunity to remain in Bulgaria working a good job, they both insisted, they would never have left. While the Bulgarian economic reality[60] does present ripe opportunities for capitalizing on individuals' desire to escape poverty through migration for work, it also calls into question a straightforward identification of source countries with vulnerable, poor, innocent "victims" exploited on their way to a better life.

It seems to me that whether negotiating the gaze of the state, a therapist, one's community, or an academic, freedom "is at odds with security,"[61] and symptoms of political distress are dismissed. This precludes meaningful conversation about "that which symptomizes deep political distress in a culture; injury is thereby rendered intentional and individual, politics is reduced to punishment, and justice is equated with such punishment on the one hand and with the courts on the other."[62] To relate this back to trafficking, the criminalization of the act of trafficking and the redresses envisioned in a new universe of punitive and corrective legal measures lead to obfuscating the ambiguity inherent in attempts to erase insecurity. Internalizing the blame for a failure to live by social norms, rather than recognizing it as a symptom of what Wendy Brown calls "social hurt," is exacerbated when "security functions by explicitly securing the self/us and implicitly relegating others to spaces of abjection."[63] To my questions regarding the stories of each of the survivors I met, I received varying answers. They were complex, intricate, contained omissions, and did not always make sense, but they were nonetheless irreducible to the label "Survivor of Trafficking in Human Beings." In some ways, this is how they went:

i want my mom
even though she made me marry
against my will

i don't have any information

but isn't the baby beautiful?

***

I am doing things that overwhelm me.
They give me choices and I say nothing
they want to train me, and I want to sleep.

In Sweden, I used to dance in front of the store
to fight off the cold.
I was raped before I was trafficked.
Different people, different times
it happened outside my house.

Hardships, struggles—
I write love poetry

***

Prostitution is amoral, dirty, and repulsive:
why would anyone want to do that?

When I realized what I would be doing
I asked to call my brother
some pretext
and told him about it and where I thought I was
I felt better, knowing the police were coming.

But they did not come.

So I jumped in front
of a moving police car
shouting for help:

Truth is on my side.

But what if, because "security practices have constitutive effects in terms of subjectivity and political effects in terms of the constitution and reproduction of political communities,"[64] the survivors "continue to be risky bodies, abject carriers of an unmaterialized threat" in which, even if "their resistance is rendered meaningless, read only as a pathological reaction rather than a political statement,"[65] disordering the social? Wendy Brown makes a similar argument in *Walled States: Waning Sovereignty*, as does Étienne Balibar in *Politics and the Other Scene*, as does Roxanne Doty in *Anti-Immigrantisms in Western Democracies* and the list goes on. As long as the trafficking survivor lays bare the inherent inability of the nation-state to address the needs and provide for the security of its subjects, s/he is dangerous, risky, and compromising. Getting away from being insecure and causing insecurity at the same time offers a lens into the shared experiences of insecurity that state both experience and create themselves.

FIXING THE BODY: FIXING THE STATE

If Cvetan, the director of a trafficking and domestic violence shelter in Bulgaria, is right that what survivors of trafficking need the most, after their daily physical needs are met, is to be heard, then what does "listening and hearing" look like if a search for the truth of trafficking, as discussed earlier, has been rejected?[66] Claudia Aradau, in her *Rethinking Trafficking in Women*, presents a simple yet shocking insight when she says, "Security discussions are present in relation to the problem of trafficking, yet absent from most of the literature on trafficking,"[67] depoliticizing trafficking and relegating it to already existing discourses of other similarly illegal border crossing activities: smuggling, prostitution, and migration. To my question about what makes a trafficking survivor secure, this is what I was told:

in the morning.
I wake up and cook an egg,
or don't,
or stay in bed until noon.
but if I wanted to cook an egg
I could, and I do,
because I can and before
I could not,

always amazed
at the hidden powers of domesticity.

I met Cvetan in his home, around 8:30 a.m. after arriving on the overnight train from Sofia. The sand on the beach was cold, fresh, and hard. I watched the sunrise and the seagulls pick seaweed and sunflower seed shells off the garbage piles. A woman was swimming, popping in and out of the cold water, while a dog quietly watched her. Backpack stuffed full of papers and an extra coat, and after a cup of coffee at a small local bakery, I walked to his apartment. We spend the next twelve hours drinking coffee and talking at his dinner table. Like with trafficking survivors, those working with them have their own stories, and often what they disclose is controversial, partial, contradictory, and sometimes even upsetting. I did not take many notes or record our conversation. But it went something like this:

8:30 a.m.: There are four tomato plants on the balcony,
a small dog amuses itself at my feet.

10 a.m.: A Chinese woman's been lying
And a Romanian's story doesn't check out
But sometimes it goes like that.

12:30 p.m.: Romas are selling their children
And their King, he is feared,
But we met, and I survived
The fear.

2:00 p.m.: Paperwork, proof of identity, documentation
All the way to Nepal
Can you help? Help!
I couldn't.

4:30 p.m.: Let's go meet some women.
Gypsy dancing. Laughter. Five cigarettes.
The numbers don't show anything

3:00 p.m.: Almost forgot—Bulgaria's been doing great
The numbers show it,
Better than the Czechs and the Romanians,

Even though your government doesn't get it,
But EU pressure brings clarity, thank God.

5:00 p.m.: Aren't you going to write anything down?
I do. My impressions of the sea.

7:45 p.m.: Dinner's ready—spaghetti—home cooked.

10:30 p.m.: The train, again! A clean, warm bed.
10:47 p.m.: Departure. Cross-country dreaming.

With reference to a pathologized, medical vocabulary of disease, illness, injury, and recovery, the retelling of trafficking experiences can be read as exposing pathological behaviors, both on the part of the survivor and on the part of the perpetrator (pimp, trafficker, madam, smuggler, client) defined as criminal, deviant, and violent. In effect, the audience outlives the accusation of being an unsympathetic bystander, empathizing and sympathizing as it is with the experiences disclosed, remaining exempt from responsibility. In this case, Todres's research on trafficking, law, and "othering" is relevant in that it provides a framework from which to understand, though not necessarily excuse, the underlying factors in the objectification and subsequent abuse of trafficking survivors. He argues that "the collective sense of the Self and the Self's perception of the Other are then given voice through the apparatus of the State. That is, the Self/Other distinction is adopted by the State and serves to shape the law it promulgates, and that law, in turn, serves to further 'affect social valuations in general.'"[68]

Allowing ethnic Bulgarians to prop their claims to lawful citizenship in which "bad actors" constructed as Others are seen as distant and "unconnected to, or in any way of shaped by, actions of the dominant group,"[69] is an example of what Todres calls attitudes toward the Other "that reflect an entrenched cultural othering."[70] That is why, when I asked an employee of a Bulgarian NGO working with survivors of domestic, sexual, and transborder violence what she thought the underlying problem was, she unequivocally said that the biggest culprits were the prejudices, attitudes, and emotional immaturity of the general public and responsible officials in Bulgaria when it came to understanding, addressing, and empathizing with survivors of any kind of sexual violence and/or other systematic exploitation,[71] which bears too much resemblance to other traumatic experiences that trouble, confuse, and unnerve. Trafficking survivors are thus, invariably and inescapably, different, outsiders, and even dangerous bodies, "second, third, or fourth category people," because they point to an inherent dysfunctionality plaguing the nation-state, a dysfunctionality that is readily explained away through stories of suffering, trauma, and pain. To my question regarding the origins of this immaturity and the consequences stemming from it, one therapist replied:

> There are laws, of course, that address the issue—but do you know how these laws are carried out? We don't have a single case won in seventeen years. Their lives are hard, their lives are difficult, they don't have work, their lives are not like your life and my life. When they come back to Bulgaria, to their families, they need support from their families. They can only fix their lives if they have a strong family environment, but the support there is missing, so they turn to different institutions. They skip from one institution to another, it is hard for them to hold a job, and the dealers look for them again. [72]

Perhaps the most provocative thing she shared with me, in the elegantly decorated, welcoming "treatment room" where the interview was conducted, was her observation that in the case of trafficking, everyone, including herself, is a little afraid of survivors, what they have lived through, and what they know. And so "we tell ourselves, this cannot be true, they must be lying, exaggerating."[73] In this sense, the violence inflicted by others becomes a perfect distraction from the fear and violence lived through as human, as trauma and suffering are invoked for their shocking, exceptional, overpowering effects. The demos is mum with shock and with complacency, while the nation-state toolbox grows and grows away from any meaningful engagement with the underlying causes of the insecurity they have helped usher.

This person who connected me
To this job
Is going to get me another.
And I am going to make a lot of money
Washing dishes

. . .
The cops, they do their job
Let me explain how it is
Let's say I am a social worker
We go to the police
Interrogation
And it is exactly like this
Question after question
Like artillery fire
And my role is to smile
And interject
So she can give good information

Because the cop doesn't even know
He is attacking her
. . .
Lack of emotional maturity

The judge is not a bad person
But he involuntarily ruined the investigation

. . .
So what do we do?
Decoupage
Do it yourself projects.
They make small trinkets, we sell them
Because research says work structures life
Because life must be goal-oriented.

The focus when dealing with survivors of trafficking and with trafficking itself always falls on the creation of boundaries, categories, and legitimate governmental interventions. Claudia Aradau uncovers an interesting problematic in which even managed and legitimate engagements by the state "create a managed space where knowledge plays an important role for the representation of trafficking" and where trafficking, the subject of a governmental apparatus responsible for addressing it, solving it, and eradicating it, promotes "the creation of a regime of truth" that is "structured and categorized in security terms."[74]

The result: a preoccupation with the body, its excesses, experiences, and ability to faithfully narrate an event, really a series of events, so that the latter can be made to fit within a victim-centered framework created by various professionals and their army of helpers. The state's job, in turn, is to erase that excess. In the Bulgarian example, the police and criminal justice system are often seen not only as incapable but also as unwilling to facilitate bringing perpetrators to justice. On several occasions during my interviews with social workers and legal professionals in the field of trafficking in human beings in Bulgaria, I was told a story involving a high-ranking justice or police official diminishing, dismissing, or altogether criminalizing survivors themselves, demonstrating an utter lack of understanding and appreciation for their position in deciding on life as a trafficked person. For example, when asked to comment on the "period for reflection," currently set at thirty days, during which time a survivor is provided with all services needed to decide whether to press charges and cooperate with the authorities in a legal case, a member of the High Court responded thus: "How long are we going to wait for her to make up her mind, why's she dawdling? What's to think about for 30 days?"[75]

Even though the "period for reflection" is hugely important in deciding whether a woman is categorized as a victim or a criminal, whether her human rights abuse claims are validated or denied, tying her to a fixed legal apparatus responsible for underwriting security not only "influences what can be said and done about trafficking" but also alternative modes of addressing trafficking, namely human rights approaches, which "become part of the interventions to deal with the phenomenon of trafficking."[76] As a result, trafficking survivors exist in multiple universes depending on who is intervening on their behalf and are defined with relation to multiple already existing categories of insecurity as reference points: illegal migrants, victims of sexual violence, smuggling cases, and prostitutes.

Even though security is not engaged as a central topic when discussing trafficking, I argue it is the ever-present organizing principle of the trafficking discourse, in which "even if human trafficking does not satisfy the conditions of possibility for a threat . . . it can be integrated into a loose definition of national security."[77] In this sense, the disempowered are further deprived of a voice. Their unsuccessful speech acts raise questions about our shared condition of epistemological insecurity in which "the pervasiveness of security practices is due to state failure"[78] even as "signs of state inadequacy in other domains become transfigured in the realm of criminal punishment."[79] The contingency, indeterminacy, and unpredictability of the risks plaguing the modern human, risks created by modernity itself, are thus calculated and whisked away in a move to erase ambivalence and "feign control over the uncontrollable—in politics, law, science, technology, economy and everyday life"[80]—where I stand with Michael Dillon in arguing that practices of security are fundamentally intended to mask, hide, and obscure the insecurity inherent in state sovereignty as such.[81] If "state security is not synonymous with the security of everyone living in the state,"[82] then securing trafficking survivors cannot be addressed without understanding the violence that plagues all political communities. In this sense, the migrant, trafficking survivor, and the refugee as subjects of security reject the exceptional in its definition as abjection.

To return to Agamben here for a bit, his framework of the exception, while a hugely useful tool for understanding the relationship between the juridical and political dimensions of modernity, does not promote resistance. While none of the trafficking survivors I spoke with were grateful for their experiences, as one may be grateful, in hindsight, for the experience of being rejected for a job, neither did they lament their lot as most had aspirations for their lives that were very closely matched to the aspiration of other Bulgarian women and men their age. The most recurring theme was the desire to have a comfortable home, a family, a job, and financial security. Yet survivors are often understood as a group of individuals afflicted with any or a combination of the following symptoms: poverty, youth, gender, gullibility, and geographic, ethnic, and demographic misfortune. This perfunctory inventory of risk factors does not necessarily address the personal experiences of victims.[83] The situation is further "remedied" through the work of expertly positioned, resource-rich states whose counseling programs, nongovernmental initiatives, national commissions on trafficking, government-level undertakings, and various international instruments for recording, studying, and raising awareness about trafficking abound.

Even as I agree with Agamben that life always finds itself "in the most intimate relationship with sovereignty,"[84] I disagree that this is because of the exception or because the relationship is a mutually and positively constitutive one. In fact, the singularity of every breathing, speaking, moving human being demands a response worthy of that singularity. This response is missing from Agamben's conceptualization and where freedom can be thought outside the sovereign and beyond security.

Judith Butler discusses "precarious life" as the "precarious life of the Other" before addressing the precariousness of life in general, insisting that "dominant forms of representation can and must be disrupted for something about the precariousness of life to be apprehended."[85] This precariousness, in the case of trafficking survivors, is directly related to dominant conceptions of the border, lawfulness, and belonging, and can best be understood when disrupted from within its own logic. In other words, if nation-states see themselves and are tasked as the primary actors responsible for "saving" trafficked human beings from violent and exploitative criminals and criminal networks, what are the implications this has for ways in which states define, lay claim to, and defend their own ability to execute this difficult task? How do these claims perpetuate a mitigation of risk dependent on regimes of exclusion, militarization of border control, and the labeling of citizens as risk-carrying Others? In Butler's register, the creation and justification of "ungrievable humans" is one of those instruments and goes a long way toward legitimizing not only the state's abilities to exclude but also our lack of empathy. Yet what is the political relevance of mourning?[86] "So it is not that mourning is the goal of politics, but that without the capacity to mourn, we lose that keener sense of life we need in order to oppose violence."[87] Is denying the opportunity to mourn the dead also a way of precluding political resistance? Or is it "primarily a discourse of the state, which provides an official interpretation of public loss in order to support a particular version of civic identity?"[88] As discussed in chapter 6, mourning has long been sanctioned by the state, especially in the context of wars and war casualties.

If we take seriously the claim that "the state is directly and inescapably the cause of vulnerability," then the victim of trafficking is "caught in an iron cage of logic, a state created subject with no room for manoeuvre."[89] In this way, and relating this back to the question of borders, "immigration controls produce relations of domination and subordination, thereby leaving state responsibility for the consequences of this completely out of the picture."[90] Interestingly enough, in addition to expanding the regulatory functions of the state, "the call for state responsibility for preventing and treating victims [is] in direct contrast to the new ethics of personal responsibility that [is] the cornerstone of the neoliberal agenda"[91] in which, alongside discourses on sexual violence in general, "the vulnerability of the victim may reflect her own failure to follow the social rules that maintain the sexual economy of conventional families."[92] In this sense, how we understand trafficking is of utmost importance not only for the public understanding of this event but also for how professionals equip themselves for dealing with it.

Importantly, the ability to cross international borders freely in the context of the EU is at the heart of the logic informing EU dynamic—in which not only goods but also people are the supposed and real beneficiaries of an open-border Europe. Trafficking and trafficking survivors, in this sense, bring the border back into focus and demand that it become one of the important and indispensable tools for analyzing trafficking itself. The victim, thus, is the woman, the city, the country: the country as a victim of historical and current exploitative market, political, and European policies; the city as a victim of being in a part of the country that "breeds" trafficking; and the woman as a victim of a collective sense of abandonment that plagues countries in regions susceptible to trafficking. This is exacerbated when the survivor is an ethnic minority.

ELLA

While in Bulgaria, I spoke with a young, underage Roma trafficking survivor who had been the focus of an anti-trafficking campaign by an NGO in the country a few years earlier. Ironic as it may be, the young woman ended up being trafficked abroad, where she underwent sexual exploitation, ended up in an Italian hospital, was rescued by a Bulgarian representative of an international NGO, and was then taken back to her parents' house where I met her, pregnant with her first child, the pregnancy not the result of sexual exploitation at the hands of traffickers. Her story, while not atypical of trafficking survivors, offers some interesting parallels when discussing the involvement of the state and the construction of victimhood that are part and parcel of the trafficking discourse.

I shall name her Ella, for the purposes of recounting her story and protecting her anonymity. Ella is a young Roma woman living with her mother and father in a bedroom that is one of three in the family's small home. The other two bedrooms are shared by her two older brothers and their wives, being that the father of sons is obligated to provide a home for his son's families. The living room/bedroom is clean though very sparingly furnished. The family does not have a working refrigerator. Ella and her mother were waiting for me to arrive. The mother had taken the day off from her agricultural job several villages away in order to talk to me and be there during her daughter's interview.

To understand Ella, one has to engage her parents. The father bragged that he was a man who treated women better than other Roma men, that he did not ask his wife to wash his feet as is the tradition and that he had found Jesus a few years earlier, leading him to give up his ways of stealing, contraband, and wood poaching. He was now the pastor of a small church, sponsored by a Swiss organization, situated in the back yard of their small home. He was now highly respected, was very proud of the fact that even though his daughter could prostitute herself abroad to get money to build a big house, examples of which lined some of the streets around him, he would not allow his blood to be insulted so. He answered most of my questions and allowed his daughter and his wife very little say, even when I directly turned to them for information. He was unable to work in the fields due to an injury to his back and was thus staying home most of the time taking care of his small granddaughter, about six months old, who was the child of one of his sons, and also of the affairs of the church, which brought him pride and a sense of importance.

His daughter was subdued, complained of pains because of the pregnancy, which was in its seventh month when I spoke to her, of not having her own room to share with the baby, of missing the father of the baby, who was working abroad in the Netherlands, and of being bored and not having anything fun to do. She was focused on her experiences in Italy only to the extent that, she said, it meant that she was in need of help right now. Her position prior to leaving for Italy was not something we discussed, but it was obvious that it was the direct cause of her departure. Her status as an ethnic minority in Bulgaria, combined with her underage pregnancy, combined with her trafficking survivor label, combined with her socioeconomic hardship and dependency, all led her to see herself, represent herself, and understand herself as someone who was a victim, and in multiple ways.

Innocence, in this sense, is understood through a perceived reality of poverty, deprivation, and victimhood and something that is reclaimed and mediated by the government apparatus. While state-sponsored programs and interventions bring trafficking survivors out of hiding and assist them as they transition to "normal" life, the tendency to turn to the state for protection rather than for critique is so prevalent as to become dangerous,[93] and in the case of trafficking in Bulgaria and other Eastern European countries creates a situation in which the legitimacy of the state is taken for granted and underwritten by trafficking itself.

Poverty, difficulty meeting their daily needs, family demands, and a desire to fulfill an idea of what it means to be a desired, comfortable, and successful person all contributed to survivors' decision to seek employment abroad, even if that means being hooked on a promise to wash dishes for thirty-five euro an hour, room and board provided free of charge—a ludicrous and unrealistic promise that should have signaled a huge red flag to any rational individual. Not only did it not do that, it came hand in hand with an idealized version of the West, the world outside: a version that overlooked the possible risks inherent in a decision to flee, rife with lack of information, inexperience, monolinguistic hell, and vulnerability. The stories of Peter and

Ella both underscore the hidden vulnerabilities that poor and developing countries face when asked to care for people who have undergone trauma, like trafficking survivors.

All trafficking survivors had difficulty in answering the following question: "How are you spending your time right now? How are you filling your days?" Ella was waiting for her child to be born, who, at the time of writing this, is already a year and a half old; Peter was waiting in a hospice for something to happen; Irina was waiting for a good job opportunity while making money selling trinkets at one of the flea markets in Sofia; Maria was expecting her first child as well, while also working at a news stand in a small provincial town; Raya was making plans to buy a house with money she did not have, working and having tea with friends who thought her place was "amazing, homely" and so well kept; the two ladies I met in a seaside town, Ivana and Stella, were waiting in a safe house for justice for those who had harmed them, doing odd jobs, and Ivana was also finishing her high school at night while trying to recover her children from the state; Elena was a temporary employee for a municipality's park services pulling weeds along the side of roads and working on beautification projects, hoping to have a child soon with her boyfriend; and the Guinean woman I met was waiting in a shelter in Sofia with her six-month-old daughter for some decision on her status as a refugee/trafficking survivor so she could figure out how to survive in a largely white, inhospitable, and difficult to understand and navigate country. She also mentioned she did not have summer shoes and asked if I could get her some.

Yet they also had another reality. Irina spoke of decorating the small apartment she shared with another survivor, of taking pains to make it comfortable, scrub the bath, buy small trinkets that would add charm to their place, and being particularly annoyed that the hot water had stopped working, though she had managed to invent a bathing solution despite this obstacle. She spent time caring for her appearance, though as a middle-aged woman, one would think these concerns should be foreign to her. She was kind, wrote poetry, was coquettish in the most endearing of ways, and worried that I would not approve of her appearance. She had a girly laugh, and every time we met she joked that she could be my mother (she could have been) and wondered how we ended up crossing paths. She was superstitious but also educated; believed in fate but also worked hard to shape her own destiny; understood that she had been exploited but did not consider herself a victim.

Where Butler is at her best is in suggesting that the practice of using one's voice to express an opinion or criticism leads to precariousness, connecting the facelessness/voicelessness of the Other to desensitization, violence, and a fabricated sense of security. To be a "precarious life" also means to tread the line of what is deemed speakable by the censorship apparatus that be, but is also "a way of establishing whose lives can be marked as / lives, and whose deaths will count as deaths."[94] Because the processes of censorship that Butler refers to in *Precarious Life* are not static or fixed but unfold according to a logic and an imperative of "false security," they are also reversible. Grief and loss, in this sense, are public and politically motivating/motivated and, as such, offer the possibility for a restorative engagement with painful, traumatic memories and past experiences. "Let's face it. We are undone by each other. And if we're not, we're missing something."[95]

Butler's "precarious life" carries an ethical charge in the way that Agamben's *homo sacer* does not—the former's insecurity derived from a status as an ungrievable, unmournable human, which adds a dimension beyond legal parlance to our understanding of the evolution of the category of *homo sacer*. "Precarious life" is different from *homo sacer* in that the former, for Butler, exists in a symbolic universe combating unknown threats with real violence, and where the lives and fates of "precarious humans," newly defined as deviant, unpredictable, and thus dangerous, can now be taken out of the public sphere as an object of debate and conversation, precluding a vigorous self-critique by the very masters of retaliatory violence. *Homo sacer*, on the other hand, stops short of eliciting an outrage and thus the possibility to resist an enemy that not only affirms the sovereign's grip on power but also creates the conditions for the enunciation of a real, breathing political singularity derived from conditions of vulnerability, dispossession, and negation. In this sense, the body implies "mortality, vulnerability, agency: the skin and the flesh expose us to the gaze of others, but also to touch, and to violence, and bodies put us at risk of becoming the agency and instrument of all these as well."[96] Thus, neither victims nor perpetrators can deny their corporeality.

CONCLUSION

I have argued that legitimate nation-states often create the conditions for the proliferation of trafficking in the context of Eastern Europe. In the process of rehabilitating, reintegrating, and empowering trafficking survivors, the central powers of said sovereign apparatus not only legitimate their claim to power but also create the conditions for the dissemination of a narrative centered on their own vulnerability, insecurity, and fear, which allows them to employ tactics of securitization in perpetua. In other words, political institutions are not necessarily repressive, and in the context of trafficking do not necessarily actively perpetuate physical violence against whole groups of people. Yet said political institutions perpetuate a certain narrative of intervention predicated on their ability to employ the legal mechanisms of state agencies and task forces in an effort to rescue "victims" without also taking responsibility for their hand in creating said "victims" in the first place. The bypassing of responsibility thus also legitimates their positions within the larger EU community, whose interventionist and criminalizing logic bestows upon member states the ability to liberate "suffering individuals" and apprehend "evildoers."

As I have tried to show in this chapter, the various ways in which survivors of trafficking in human beings are understood and addressed, as well as managed and governed, by the state, social workers, nongovernmental institutions, the legal system, and academics alike, hardly present an apparatus dedicated to recovering their agency. Processes of pathologization and securitization are supplemented by locating the violence and trauma suffered by an individual within discourses on risk and insecurity. The survivor of trafficking in his/her vulnerability stands in for the vulnerability and insecurity that fundamentally define the nation-state itself, an insecurity that is reaffirmed with recourse to strengthening state jurisdiction over territorial borders and thus representing all migration as an "invasion syndrome,"[97] all trafficking as risk, and all risk in turn as insecurity. As a result, trafficking is not only a legal but also a political category,[98] and as such often hides state violence done to "victims" but also uncovers opportunities to undo said violence. My engagement with the deaths of children as a result of violence conflict in chapter 6 will conclude and further inform this resistance.

NOTES

1. Kelly Oliver, *Witnessing: Beyond Recognition* (Minneapolis: University of Minnesota Press, 2001), 7.
2. Paul Kottman, "Introduction," in *For More Than One Voice: Toward a Philosophy of Vocal Expression*, ed. Adriana Cavarrero (Stanford, CA: Stanford University Press, 2005), xxi.
3. Ibid., xxiii.
4. Étienne Balibar, *Politics and the Other Scene* (London: Verso, 2002), 3–4.
5. The poems in this chapter came about by combining testimonies from different survivors of trafficking. When deemed appropriate by the author, words have been changed to fit the poem, but the overall meaning and the bulk of the actual words come directly from survivors' testimonies as they were told to me.
6. Author interview, village in central Bulgaria, April 26, 2017.
7. Giorgio Hadi Curti, "From a Wall of Bodies to a Body of Walls: Politics of Affect/Politics of Memory/Politics of War," *Emotion, Space and Society* 1 (2008): 107.
8. Ibid.
9. Marita Sturken, *Tangled Memories: The Vietnam War, the AIDS Epidemic, and the Politics of Remembering* (Oakland, CA: University of California Press, 2007), 9.
10. Curti, "From a Wall of Bodies to a Body of Walls," 108.
11. Ibid.
12. Author interview, Sofia, Bulgaria, May 4, 2017.
13. Zygmunt Bauman, *Wasted Lives: Modernity and Its Outcasts* (London: Polity, 2004).
14. Balibar, *Politics and the Other Scene*, xi.
15. Bauman, *Wasted Lives*, 4.
16. Ibid., 6.
17. Ibid., 5.
18. Ibid., 40.
19. Ibid., 41.
20. Ibid., 7.

21. Wendy Brown, *States of Injury* (Princeton, NJ: Princeton University Press, 1995), 17.
22. Bauman, *Wasted Lives*, 45.
23. Ibid., 83; emphasis in original.
24. Giorgio Agamben, *Homo Sacer: Sovereign Power and Bare Life* (Stanford, CA: Stanford University Press, 1998); Giorgio Agamben, *State of Exception* (Chicago: University of Chicago Press, 2005).
25. Oliver, *Witnessing*, 63.
26. Judith Butler, *Precarious Life: The Powers of Mourning and Violence* (London: Verso, 2004).
27. Shoshana Felman and Dori Laub, *Testimony: Crisis of Witnessing in Literature, Psychoanalysis and History* (New York: Routledge, 1992), 53.
28. Ibid., 5.
29. Butler, *Precarious Life*, xx.
30. Brown, *States of Injury*, 75.
31. Balibar, *Politics and the Other Scene*, 35.
32. Mariella Pandolfi, "From Paradox to Paradigm: The Permanent State of Emergency in the Balkans," in *Contemporary States of Emergency: The Politics of Military and Humanitarian Interventions*, ed. Didier Fassin and Mariella Pandolfi (New York: Zone Books, 2010), 156.
33. Ibid.
34. Ibid., 159.
35. Ibid.
36. Ibid.
37. Ibid., 162.
38. Kottman, "Introduction," xxi.
39. Author interview on April 28, 2017.
40. Thanks to my former student and friend Rikki Boucher for pointing out this relationship.
41. Bauman, *Wasted Lives*.
42. Jo Doezema, "Loose Women or Lost Women? The Re-Emergence of the Myth of White Slavery in Contemporary Discourses on Trafficking in Women," *Gender Issues* (Winter 2000): 23–50.
43. Felman and Laub, *Testimony*, 20.
44. Vicky Squire, *Post/Humanitarian Border Politics Between Mexico and the US* (London: Palgrave McMillan, 2015), 31.
45. Butler, *Precarious Life*.
46. Agamben, *Homo Sacer*, 142.
47. Dolores Chew, "Gender, Migration and Trafficking—An Introduction," *Labour, Capital and Society* 39, no. 2 (2006): 8.
48. Maggie McNeill, "Mid-Witness Testimony: The Unreliability of First Person Accounts in Sex Trafficking Discourse," *Albany Government Law Review* 7 (2014): 87.
49. Elizabeth Bernstein, "Militarized Humanitarianism Meets Carceral Feminism: The Politics of Sex, Rights, and Freedom in Contemporary Antitrafficking Campaigns," *Journal of Women in Culture and Society* 36, no. 1 (2010): 54.
50. Kristin Bumiller, *In an Abusive State* (Durham, NC: Duke University Press, 2008), 13.
51. Ibid., 20.
52. Doezema, "Loose Women or Lost Women?" 41.
53. Bumiller, *In an Abusive State*, 129.
54. Rutvica Andrijasevic, "Beautiful Dead Bodies: Gender, Migration and Representation in Anti-Trafficking Campaigns," *Feminist Review* 86 (2007): 24–44; Doezema, "Loose Women or Lost Women?"; Jan Nijboer and Judith Vocks, "The Promised Land: A Study of Trafficking in Women from Central and Eastern Europe to the Netherlands," *European Journal on Criminal Policy and Research* 8 (2000): 379–88.
55. Bumiller, *In an Abusive State*, 131.
56. Jo Doezema, "Ouch! Western Feminists' 'Wounded Attachment' to the 'Third World Prostitute,'" *Feminist Review* 67 (2001): 16–38.
57. Bumiller, *In an Abusive State*, 31.
58. Ibid., 34.

59. Author interviews, Sofia, Bulgaria, April 28, 2017, and June 10, 2017.

60. According to an article published in November 2016, "Bulgaria has highest percentage in the European Union (EU) of people at risk of poverty or social exclusion—41.3%, media giant Bertelsmann says in a report. One-third of the Bulgarians have no resources to cover basic needs, necessary for normal life, according to the report. In addition, in two of six dimensions of social justice, our country ranks last. These are the prevention of poverty, social inclusion and non-discrimination. Meanwhile, Bulgaria is among the countries that set aside the lowest amount of funding for science and innovation. There is a large proportion of people who neither study nor work in Bulgaria. The report said, that according to last year's report 24% of young people aged between 20–24 years are not engaged with anything. In the EU the average level is about 17.3% in the group of 20–24 years old, in Italy—31.1%, Greece 26.1%, while in Spain the percentage is 22. Bertelsmann made also a few statements according to which 34.2% of people in Bulgaria do not have enough money to cover their daily needs, such as heating their apartment or paying their phone bills. The highest risk is for the elderly and children. The group of retirees 51.8% are at risk of poverty, while 40.9% of them suffer from serious material deprivation"; http://www.novinite.com/articles/177427/High+Risk+of+Poverty+or+Social+Exclusion+for+People+in+Bulgaria+%E2%80%93+Report (accessed November 30, 2017).

61. Brown, *States of Injury*, 25.

62. Ibid., 27–28.

63. Claudia Aradau, *Rethinking Trafficking in Women* (London: Palgrave McMillan, 2008), 61.

64. Ibid., 8.

65. Ibid., 9.

66. Aradau, *Rethinking Trafficking in Women*; Doezema, "Loose Women or Lost Women?"; Doezema, "Ouch!"; Jo Doezema, *Sex Slaves and Discourse Masters: The Construction of Trafficking* (London: Zed Books, 2010).

67. Aradau, *Rethinking Trafficking in Women*, 16.

68. Jonathan Todres, "Law, Otherness, and Human Trafficking," *Santa Clara Law Review* (2009): 3–4.

69. Ibid., 5.

70. Ibid., 7.

71. P.U.L.S. employee, author interview, Pernik, Bulgaria, March 2, 2017.

72. Ibid.

73. Ibid.

74. Aradau, *Rethinking Trafficking in Women*, 18–20.

75. Author interview, February 6, 2017.

76. Aradau, *Rethinking Trafficking in Women*, 37–38.

77. Ibid., 43.

78. Ibid., 53.

79. William Connolly, *Why I Am Not a Secularist* (Minneapolis: University of Minnesota Press, 1999), 129.

80. Ulrich Beck, *World at Risk* (Cambridge: Polity Press, 2009), 41.

81. Michael Dillon, *Politics of Security* (London: Routledge, 1996).

82. Aradau, *Rethinking Trafficking in Women*, 57.

83. Bumiller, *In an Abusive State*, 81.

84. Agamben, *Homo Sacer*, 67.

85. Butler, *Precarious Life*, xviii.

86. David McIvor, "Bringing Ourselves to Grief: Judith Butler and the Politics of Mourning," *Political Theory* 40, no. 4 (2012): 410.

87. Butler, *Precarious Life*, xiii–xix.

88. McIvor, "Bringing Ourselves to Grief," 410.

89. Bridget Anderson, "Stuck in Traffic: How Helpful Is the Trafficking Framework," www.compass.ox.ac.uk (accessed November 12, 2017).

90. Ibid.

91. Bumiller, *In an Abusive State*, 5.

92. Ibid., 20.

93. Brown, *States of Injury.*
94. Ibid., xx–xxi.
95. Ibid., 23.
96. Ibid., 26.
97. Rutvica Andrijasevic, "The Difference Borders Make: (Il)Legality, Migration and Trafficking in Italy among Eastern European Women in Prostitution," in *Uprootings/Regroundings: Questions of Home and Migration*, ed. Sara Ahmed, Claudia Castada, Anne-Marie Fortier, and Mimi Sheller (Oxford: Berg Publishers, 2003), 254.
98. Aradau, *Rethinking Trafficking in Women.*

Chapter Five

Waiting for the State

Tibetan Refugees in Nepal

TASHI PELKHIEL TIBETAN REFUGEE SETTLEMENT, POKHARA, NEPAL

It was hot and humid when I arrived. Green Tibetan flags guided me, as they had in other settlements, and I found my destination easily. The settlement was clean, orderly, and quiet, so quiet that it was hard to believe people lived there. Next to the bus stop was a small market of tea stalls and jewelry vendors, and to the right stretched the unpaved but wide and even path toward the settlement. Then it came—the dog. He was neither big nor small. He was black, with long hair interrupted by reddish yellow spots, and had the comfortable gait of a local resident. He followed a safe distance behind and would stop every time I stopped. Once, when I was not sure which way to go, he guided me to the left, through a maze of houses and toward the center "square." His was a comforting presence, as if he had been waiting to show me the settlement and the river, where he watched me watch a man and his oxen plow, where another man washed clothes in the river, and where a grandfather was dozing under a flapping yellow tarp. The dog was not intrusive and did not make me feel unsafe, but he also would not leave. When I would go inside a building, he would lie in the dirt outside and wait until someone shooed him away, only to find me again when I exited the building to resume his escort.

The walk to the river is my fondest memory. It was high noon and hot. Because most of the kids in the school were boarding kids, there was not the usual hustle and bustle of running home from school during the lunch hour. I was taking a walk on one of the settlement streets, taking pictures, enjoying the quiet, and trying to figure out how to organize the rest of the day in the most efficient way. The dog kept pressing on, so I followed. We soon left the settlement and walked toward what seemed like a cliff going nowhere. In the distance, I could see the outlines of a monastery, its white walls hot against the midday sun. Then the road curved and became steep—garbage and strange orange rocks detracting from the scorching sun. Whenever I would stop to look around, the dog would stop too. To our left was what looked like a zip line station. I was curious and walked toward it. There was a guard at the door, an old man in his sixties or seventies, who waved me in. But not the dog, who waited down a steep set of steps for me to return and collect him. I walked inside, where there were a group of young Nepali men, boys really, playing cards and eating lunch; they waved to me but did not try to sell me a ride. The zip line went up in the surrounding hills and soon disappeared into their green thickery. I remember walking through similar hills some years earlier, salt container in my pocket, leeches stuck to my ankles, the humidity as thick as a shower. As I was not interested in zipping to and fro in the hills, and as my companion was waiting for me, I soon left. When I resumed my descent, the dog resumed its descent as well. I had injured my foot some days before and the steep descent on rocky terrain felt painful, so I made my way gingerly among empty Pepsi bottles and chips bags. Soon we were right above the river, beyond which lay several plots of agricultural land. When I stopped to sit down, the dog laid next to me, rested, then got up and trotted to the water.

A big heron-looking bird was marching in the shallow water, dressed as if for a party, looking somehow out of place among weedy plots of land, skinny cows, and strips of wiring floating around. The dog walked toward the slow-moving water, stopped as if to accustom itself to the cooler temperature, looked around, and laid down its whole body in the slow-moving water. It gave me comfort to watch him, what with the heat and the stillness, which both felt oppressive. I focused my attention on my furry companion; he laid there for what seemed like an hour, but my watch said it was only five minutes—one of my initial observations on the varied ways of passing time and of time passing around refugee settlements. Then he got up, drank some water, walked back up to me, laid down again about three feet away, and dozed off. I did not dare get up and disturb him, so I remained seated and still. In the distance, a man in shorts and tattered T-shirt was plowing a field with an ox. His high voice seemed, for a while, the only noise around, and its rhythmic tone somehow matched the ox's gait. They were well paired. Behind them, from the hills, appeared two dozen small domesticated goats. I could not discern the path they were using until I saw their organized trot downhill on what seemed like a thread-thin line in the thicket. I counted about twenty. They came down in a mad run, but as soon as they hit flat ground, they froze and laid down. There wasn't a goat herder in sight until, some ten minutes later, a young boy who could not have been older than twelve emerged from the thicket to claim them as his. He was not wearing a hat, had a stick for a third leg, and did not seem to be carrying any water or food. I took a sip from my store-bought mineral water and watched as he washed his face in the river before laying down to rest in a shady spot.

When I judged it time to resume my field work, I stood up, my canine companion in tow, and walked back to the settlement. In its midst was a small monastery chapel, its door open, several pairs of flip-flops parked outside. I could not see anything inside—it was so bright where I was standing and so lightless inside. I kept walking. I took some more pictures of the houses because, like before, I was struck by their orderliness, cleanliness, and neatness. I was reminded of walking through a Tibetan refugee settlement in Mustang in 2003 and feeling the same way. The houses were white with blue trim, and all around flapped Tibetan prayer flags. By the time I arrived back at the settlement officer's office, it was close to 1:30 p.m. We spoke again, having already done the official interview before lunch, this time about hiking, the remote settlements in the mountains he was in charge of, and the arduous task of getting there on foot or by jeep in the summer (no access in the winter). He said his knees hurt too much to do much running or hard hiking anymore, though he did not look older than thirty-five.

Part of his responsibilities as camp manager included overseeing the remote settlements under his jurisdiction, including settlements in which some of the school children's parents lived. He said the kids, because of the distance and the difficulty of getting there, did not get to see their parents for months at a time, and that parents rarely had the opportunity to come to Pokhara. Christmas break and the longer time off school at that time were one of the few opportunities kids had to be with family during the school year. To my question relating to parents' willingness to send their kids away, he responded: "They know their kids are better off here, that education is needed for their future." Education, indeed, is one of the cornerstones of His Holiness the Dalai Lama's policy for Tibetans in exile and something that was repeatedly reiterated to me throughout my visits to settlements. I thought of my seven-year-old son going to school and only seeing me during winter break—not only could I not comprehend this, but I also failed to see the utility, no matter the circumstances. Privileges.

To our left was the monastery, well known in town also for hosting a steady group of tourists on retreats. There was a multiday event underway and I was unable to go inside, though I did see, through wide-open windows, what looked like nuns doing prostrations. The heavy beat of a chant filled the midday air, thick with the heat and the moisture, still as mud. As always, during all my conversations with Tibetan Government in Exile (TGiE) officials or other Tibetans for that matter, a Tibetan woman with tea would materialize as if out of nowhere, prepared with milk, butter, and spices.

"Thuk Je Che" (Thank you), I would say, and she would be gone. After the obligatory conversation about His Holiness, lack of employment, youth not staying engaged, and Chinese policies, the settlement officer directed me to the school. I had almost forgotten about the dog, but when I looked behind me he was there, following.

The school was an L-shaped building with concrete floors and a grassy dirt field for sport activities. I counted six classrooms, one staff room, a small makeshift kitchen, and a toilet. It was recommended to me that I meet the principal, but he had gone to Pokhara for the day to run errands. Instead I sat and waited in the staff room for someone to show up and talk to me. To the right, children were reciting something in Nepali and to the left there was a tremendous ruckus, so much so that I went to see what was going on. I had become accustomed to hanging around young children, having spent the past week with the students at Atisha Primary School in the Jawalakhel Tibetan settlement in Kathmandu. Oblivious to the fact that I was, in fact, interrupting the classroom activities, I stuck my head through the doorless wall and drew all attention to myself. I was invited to come in and sat with the kids, all in late elementary ages, their discussion on something to do with life in exile—something I wanted to know more about and the main reason I had come to Nepal in the first place.

I had arrived in Nepal on a cloudless early morning in May, even though the haze from urban dust pollution was so thick that one would not have known the blue skies were indeed cloudless. After paying $25 for a tourist visa, receiving my handwritten receipt, exchanging the necessary pleasantries with the young immigration official, collecting my bags, and getting lost twice on my way out of the brick building that is Tribhuvan International Airport in Kathmandu, I was officially in Nepal. Indeed, my visa had just granted me more rights than the Tibetan refugees I was going to work with had, even if the majority of them had been born in Nepal.

After haggling with the guys out front over taxi rates and where I was actually to be taken, it was agreed that I would be driven to my guesthouse in Patan, the old city on the outskirts of Kathmandu, for 800 rupees (about $8). Driving through 6:30 a.m. traffic in Kathmandu was enough to wake one up. One would not expect the traffic jams to be quite so big, yet they were. We circumvented cows, huge piles of garbage, tea (chija) stands, the temporary domiciles of homeless people, more huge piles of trash populated by hungry cows, and countless mopeds zooming to and fro. Finally, I arrived in my guesthouse and looked out the window at the kitchen of the house next door, where a young housewife was making breakfast and a half-naked baby was entertaining itself by sticking its toes in its mouth. From the street could be heard the morning hustle, which, as I would come to understand, would begin around 6:00 a.m. every day.

EVERY STORY HAS A STORY

Every story has a beginning whose meaning only sometimes stays true to the story. Mostly, it is the product of the hard work of making stories mean one thing or another. I went to Nepal in search of a story I thought I had the beginning of and came back without the story I had been hoping to find, yet not storyless either. I looked for stories every day: recorder in hand, research notes in my bag, and a face mask to protect me from the thick dust—something about new pipelines being installed all over the Kathmandu valley and none of the roads being repaired afterward. In the process of looking for my story I was often frustrated, disappointed, and simply exasperated by the amount of obstacles I felt stood in my way, the lack of "truly novel" discoveries I felt I was making, and time constraints. Interviews were often hurried and felt irrelevant to political science: a mere listing of facts, statistics, and placations for my Western ears. Would I be able to create a narrative afterward? Would the visit materialize into fruitful findings and conclusions? Would it help tell my story?

The story, as it turns out, wasn't mine to tell at all, and the research in Nepal became an exploration into what I will call the "politics of emotion" as an important starting point for engaging with displaced populations. Questions about sovereign states, nation-state directives, and political decisions were all interpreted and answered from the lens of people's feelings, and short of having taken videos of our interviews, a presentation of refugees' testimonies in the form of poetry is the closest I can get to remaining faithful to the storytellers while also relating a story that others can understand. Suffering, pain, displacement, and deprivation carry a deep and important political charge and as such open up the possibility of stepping away from politics in the major key of nation-states and moving closer to honoring the oft forgotten or ignored reality of the emotionally charging and demanding experiences that refugees, in particular, deal with and negotiate daily for many years of their lives.

In other words, how does a specific and intentional focus on the personal and emotional experiences of refugees interrogate the status and nature of the nation-state on the world scene? Considering that being displaced from one's home is a deeply emotional experience, how does a focus on said emotional experience in all its manifestations provide an additional, deeper lens of looking at and understanding displacement? Is a "politics of emotion" also an approach that can be used when examining other experiences of people or whole populations that are traumatic in nature, especially ones that deal with crossing international borders? How are the bodies and experiences of children in/at war engaged with in the next chapter a testimony to how we can better put to work the insights that a commitment to a "politics of emotion" yield? Last but not least, what if honoring the emotional side of people's experiences with war, displacement, and trauma helped let go of the methodological burden of having to justify and seek rational engagements with all things political and, as a result, infuse some of the stuff that real life is made out of into the study of relations between and among nation-states, organizations, tribes, communities, and individuals?

***

Can you tell me about your life?

"For example, I have my daughter and two sons."

"I am from Pokhara. My parents are still there. My mother lives in settlement and my father sells souvenirs."

"There is no duality in my personality."

***

ON DEMOGRAPHY IN HISTORICAL CONTEXT

The population of Tibetan refugees in Nepal technically makes up part of the Nepali state, yet does not fully or even legally belong to it. They are neither truly refugees nor stateless people but a bit of both as they fall under the jurisdiction of the United Nations High Commissioner for Refugees (UNHCR), yet most do not hold official refugee status. The four settlements (camps) I visited, two in Kathmandu and two in and around Pokhara, were not big, nor were they separated from the outside world by camp-like fences, but they did constitute a separate world organized according to different rules and traditions. The Jawalakhel settlement in Lalitpur, Kathmandu, was the one I spent the most time at and the one that represents most of the experiences described in this chapter. I conducted interviews with various people, ranging from settlement officers in three of the four settlements I visited, to school-age children, to housewives, to teachers, to university students, to settlement personnel, to handicraft workers, to single moms, to nuns, to shop owners. I also visited the Central Tibetan Administration (CTA) office in Kathmandu, where I spoke to Migmar Dorjee, the secretary, as well as to Karma Jelseng, a settlement officer for some of the remote Himalayan areas. It should be noted that the CTA is the name used in Nepal for the TGiE—a more neutral and apolitical designation of Tibetan authorities in exile, one acceptable to the Chinese, now a close economic and political partner of the Nepali government, and my preferred term here.

A word on historical context: the Tibetan population in Nepal consists of the descendants of the original Tibetan refugees who came to Nepal after the 1959 uprising in Tibet and the subsequent violent response to it by the Chinese authorities. It was during this time that His Holiness the Dalai Lama fled Lhasa, Tibet, for Dharamsala, India, where he resides to this day and where the TGiE headquarters are located. At the time of the uprising, as many as 20,000 Tibetans fled for Nepal, and while a number of them subsequently moved to India, some 15,000 to 20,000[1] remain in twelve settlements. The exact number of Tibetans in Nepal is unclear, with some sources estimating there to be 30,000 Tibetans in the Kathmandu valley alone.[2] Besides the seven "urban" settlements in and around Kathmandu and Pokhara, the other settlements are largely in the remote Himalayan regions of Nepal. The settlements in and around the Mustang region are known for their history of resistance to the Chinese in the first fifteen years after the 1959 uprising when armed resistance, at the request of His Holiness, ended in the 1970s. Ironically, part of the appeasement process also granted refugee card (RC) or citizenship to residents of these settlements, as was evidenced to me by some of my informants.

Between 1959 and 1989, there was only a small trickle of Tibetan refugees leaving Tibet due to stricter travel restrictions imposed by the Chinese government, making travel to Nepal technically illegal.[3] The year 1989 marks the second wave of Tibetan arrivals in Nepal and India, which is currently also down to a trickle due to new travel restrictions and subsequent political and economic agreements between Nepal and China also affecting the fate of Tibetans in exile.[4] In the period from 2010 to 2011, Chinese foreign aid to Nepal amounted to $18,843,988 or 1.7 percent of all foreign aid, which stood at $1,079,710,554. From 2011 to 2012, that figure went up to $28,344,923 out of $1,045,297,273 total, and by the next year, the figure had gone up to $44,994,559 or 3.6 percent of all foreign aid. For the past two years, it has stayed around mid to upper $30,000,000.[5] Even though India's aid to Nepal has been dropping, it is still a considerable amount and more than what China gives yearly. "India's actual disbursement of aid to Nepal in the financial year 2010–11 was USD 50,728,502. In 2011–12, it was USD 50,620,749 and in 2012–13 it rose to USD 63,813,269."[6] From 2017 to 2018, Indian aid to Nepal was at $2,989,599,639.00. The previous year, Nepal received an even larger sum at $3,879,902,139.40. Bhutan, to date, remains the largest recipient of Indian aid. Strong emphasis in Indian aid to Nepal remains on money for development projects and other nonmonetary contribution in the form of "technological assistance, preferential market access, training, lines of credit and grant in assistance and human resources development" as well as a strong emphasis on grants based on need, as was the case with the Nepal earthquake relief efforts in 2015.[7]

While China has become another main donor of foreign aid, it has also engaged in various development projects, with Nepal responding in kind with respect to the Tibetan question.[8] "On 30 April 2010, the Nepali Department of Immigration (DoI) sent nine Tibetans to jail after they refused to pay fines for illegal entrance to Nepal. Nepal has supported Beijing's 'One China' policy which declares that both Taiwan and Tibet are inalienable parts of Chinese territory."[9] Nepalese policy has continued to be one of tolerance toward the Tibetan populations but also one discouraging overt political activities of any kind. More recently, in an August 2017 visit to Nepal, the Chinese Vice Premier vowed to strengthen further the bilateral economic as well as political ties between the two countries, with talks touching on things like strengthened border control through a "joint-action center" and a Kathmandu-Tibet railroad.[10] "The connection between economic investment and suppression of Tibetans was confirmed in the Chinese state media, with the Global Times stating that one of the reasons for China's increasing investment was to reward Nepal for its 'important role in guarding against Tibetan separatists.'"[11] Additionally, trade relations between Nepal and Tibet have been growing, where Nepal's trade with Tibet in 2001/2002 was at "7.9 percent . . . and further increased to 10.3 percent in 2010/11."[12] In May 2017, Nepal sought Chinese assistance for a development project in "16 mountainous and hilly districts bordering China."[13]

***

Is China pressuring Nepal?

"I hope Tibetans and Chinese will be friends one day."

"It is not as simple as this, though you are teaching political science."

"Greed is another fact."

"Poverty is another factor."

"Change is happening. They cannot control everybody."

***

Making matters worse for Tibetans in Nepal, the latter has not signed or ratified the United Nations Convention on the Rights of the Refugee[14] (either its 1951 or 1967 iterations, referred from here on as "the Convention"), which means that, technically speaking, the Nepali state is not obligated under international law to host, care for, or recognize refugees from any country. Had Nepal signed and ratified the Convention, the result may not have been much different, seeing how the Convention remains firmly lodged and given meaning through the actions of nation-states who "are required not to return asylum seekers if there is a credible chance of persecution on return, yet need not grant them full privileges of leave to remain."[15] The gentleman's agreement that exists between the UNHCR and the Nepali government is quite unique in that it grants Tibetan refugees right to asylum without any specific protections and, in addition to ensuring the safety of Tibetan refugees in Nepal, it is supposed to ensure their safe transit to India. The position of the Nepali state has been one of tolerance and hospitality, especially pre-

1989, as remarked on by all refugees I spoke with as well as by the UNHCR head of mission to Nepal. Yet naturalization is not practiced for Tibetan refugees born in Nepal. Once a Tibetan refugee, always a Tibetan refugee.

***

Where is your home?

"In Tibet."

***

I chose to study the Tibetan refugee population because of their status as protracted refugees and their precarious situation as residents of several countries. The state of being in exile is usually a temporary one in which "the purpose is to return home, not to become established in a host state."[16] I had previously worked with the Bhutanese refugee population in Nepal. Since 2007, close to 110,000 refugees (most of them Bhutanese but also some Tibetans) have been resettled "to the United States, Canada, Australia, New Zealand, Denmark, Norway, the United Kingdom and the Netherlands."[17] The plight of the Bhutanese refugees had, for many years, remained unrecognized, with close to 100,000 of them living in six different settlements in the eastern part of Nepal, near the towns of Damak and Jhapa. They had been "expelled" from the south of Bhutan by the Bhutanese king in the early 1990s due to the perceived threat they presented to the Bhutanese state, a threat perpetuated by their refusal to obey laws enacted by the government and aimed at assimilating them into mainstream Bhutanese culture against their own cultural, ethnic, and linguistic traditions. In Bhutan, "during the 1960s and 1970s, with the development of education, social services and the economy, many Southern Bhutanese rose to occupy influential positions in the bureaucracy,"[18] which led to efforts, in the late 1980s and early 1990s, to diminish their influence and consolidate power for ethnic Bhutanese only.

Even though most Southern Bhutanese, also known as Lhotshampas, were granted citizenship after the passing of the first Bhutanese citizenship act in 1958, they did not mingle with the northerners and remained a relatively isolated and self-reliant community.

> A new citizenship act passed in 1985 became the basis for a so-called census exercise in southern districts, in which every member of the southern population had to produce documentary evidence of legal residence in 1958, or else risk being declared a non-national. In 1989, all Bhutanese became liable to a fine or imprisonment if they ventured out in anything other than northern traditional costume, and the Nepali language was removed from the school curriculum.[19]

Because of these political and cultural repressions, many Southern Bhutanese took to the streets, which resulted in imprisonment, confiscation of property, the demolition of their homes, and branding them "anti-nationalist." Most of them arrived in Nepal in the early 1990s and were settled in the two southeastern provinces of the country in six main camps.

At that time, the Nepali and Bhutanese governments were in the middle of talks including UNHCR and refugee representatives that aimed at resolving the situation. India's position, that the issue was of a bilateral nature, did not help matters, and eventually seven rounds of talks ended in an impasse, which forced the international community to take action. By 2017, most Bhutanese camps had been closed and I was not allowed access to the ones remaining as no foreigner was permitted to enter the camps. The first time I had visited, I had "forgotten" to ask official permission to conduct research. Ironically, I gained access simply by virtue of making my way across a nonexistent camp border without permission, facing the authorities with a done deal and no choice but to allow me to stay: the advantage of being a twenty-three-year-old graduate student with little precise idea of what she was doing but doing it anyway. In 2017, the Nepali Home Ministry, which is in charge of refugees in the country, declined to explain the reason for not allowing me to visit, and the UNCHR head of mission in Nepal, Kevin Allen, assured me regretfully that he could not force the matter on my behalf.

Compared to the Bhutanese case, the Tibetan situation invites a careful examination of its own. I learned that close to 75 percent of the Tibetan refugees in Nepal do not possess an RC and that a good number of them do not even have a birth certificate, which makes them technically "paperless nobodies" in the political and legal sense of the word. As a result, they cannot own property or a business, purchase a car, house, or land, or open a bank account. The children of refugees with an RC technically have legal status in the country, as their names are recorded on their parents' RC, though their own children do not inherit this. Additionally, RCs are up for renewal every year, which means that with the passing away of a parent, even the names of children on the RC become "invalid." Because of the open border between India and Nepal, Tibetan refugees are technically able to travel to India freely, but everyone I spoke with said they could only do that with ease when willing to bribe the border officials (the amount I was quoted was anywhere between Rs100 to 300, or between \$1 and \$3). My friend Tashi described the Nepali people, on the other hand, as kind, empathetic, tolerant, and hospitable. "Nepali are nice people. Law is not. Nepali government is not good, not helpful, takes bribes, corruption, especially recently."

Traveling to a country other than India, however, is not easy, and unless authorized by the Nepali government, refugees cannot resettle in a third country under the auspices of UNHCR, which was the case in 2001–2002 when the United States invited one thousand Tibetan refugees for resettlement and the Nepali government refused to issue them travel documents. This is the case because they are considered the beneficiaries of "firm settlement" in Nepal, precluding them from resettlement and third-country asylum.[20] Additionally, the original settlements built on land purchased by the Nepali Red Cross have not been expanded, and overcrowding, as one settlement officer in Kathmandu put it to me, was one of their main problems. Because of that, more recent arrivals live scattered around Kathmandu and Pokhara, as settlements cannot accommodate them. As a result, these refugees are harder to reach when they need attention. What is more, as families have grown without the appropriate housing expansions, many Tibetan homes house three generations in the space that was allotted to a single family in 1959.

Most Tibetan refugees I spoke with indicated that they felt "freer" in Nepal compared to how they would have felt were they in Tibet, despite the latter's economic coming of age at the hands of the Chinese and Nepal's largely unstable political system seeing the coming and going of twenty-six governments in the past twenty-three years. To quote the science teacher in one of the schools I visited, "I am lucky to live in a peaceful country."[21] Still, the refugees' most prevalent hopes were their desire to return "home" to Tibet, alongside wishes for His Holiness the Dalai Lama's long life, finding a good job, finding a stable partner, and having enough money to survive.

***

Do you feel at home in Nepal?

"I feel at home here but it's not homely. Insecurity is there."

***

It is important to situate the discussion of Tibetan settlements in Nepal in the context of His Holiness the Dalai Lama's teachings of peace, toleration, and nonviolence. In fact, this attitude of pacifism was quoted by many as the main reason Tibetans had hope that their problem would be resolved and that the Chinese would give in to their demands for an autonomous Tibetan territory on some part of current Tibet. As one of His Holiness the Dalai Lama's speeches recounts it, "We are all Tibetan. We have one special problem. We don't have our freedom. We need more education. We cannot fight the Chinese with a military so we fight with our brain."[22] As his faithful followers, Tibetan refugees insist on situating their behavior, dreams, and beliefs within the teachings of His Holiness, whose leadership, even if from afar, sustains the refugees in everything they do. There is a very strong emotional connection not only to the actual teachings of His Holiness the Dalai Lama but also to his persona, to the ideas and dreams he epitomizes. It would be hard to describe this as the same connection regular citizens feel to their governments, but in their intensity and unfailing devotion, I argue that their dedication to His Holiness's teachings and vision for all Tibetans exceed any known political commitment we in the West may be familiar with through our civil, political, and other engagements. In a word, His Holiness is family, he is a father to all. He exists in people's imaginations but also in their daily interactions. There is no term that can explain this special emotional connection that, for the purposes of this chapter, imbues the relationship Tibetans have to His Holiness with extreme political significance for how we create community and how we sustain one in the midst of insecurity.

I cannot remember an occasion during my interviews when His Holiness's name was not invoked in response to a question regarding "coping" or "the meaning of remaining Tibetan" in Nepal. So unanimous was everyone's agreement on the importance of his teachings and philosophy that, after a while, I stopped asking about him altogether, assuming he would come up inevitably. It was, and still is, difficult to rationalize this kind of devotion to a man who has been at the helm of the Tibetan community for decades and whose work has not yielded concrete solutions to their situation, though he has been instrumental in bringing attention to the Tibetan cause and securing donations to cover operating costs for the TGiE, scholarships for Tibetan children, and the setting up of the political arm of Tibetans in exile. The scientific need to understand the Dalai Lama vis-à-vis the nation-state presents a problem because neither the former nor the latter relate to each other in ways we are used to conceptualizing leadership and national government. This evokes the need to treat the Tibetan case with an eye to how it is different and not so much trying to align it with studies and discussions of state parties' actions and rationales for existence. It is this invitation to welcome an alternative framework for positioning the state vis-à-vis both its citizens and danger that will inform the remainder of this chapter. And it is with reference to His Holiness's work and vision that I first thought of my research as an example of the work of a "politics of emotion."

***

Why is the Dalai Lama so revered?

"Kindness, he is a very good leader."

"He is preaching love and peace."

"He is honest and always honest."

"He is our father."

"The Middle way."

***

***

Are you afraid of what will happen when the Dalai Lama passes?

"I am worried."

"If he is not here, a vacuum will be created."

"He has said that he will live more than 100 years."

"The political system is going well, so even when His Holiness passes, then I think it is really not such a huge problem."

***

The "Middle Way" policy envisioned by His Holiness and his admonition against violence and armed resistance both presuppose an expectation on the part of the TGiE that the Chinese government would come to the table and compromise in an acceptable way. This has not been the case thus far, nor does it appear that the Chinese intend to allow the return of Tibetan refugees/dissidents, coupled with the free practice of their religion, observation of their cultural, religious, and economic ways of life and speaking of Tibetan language. To untangle the conundrum that this reality presents is not the goal of this chapter. What I do address, however, is the condition of "statelessness" that the Tibetan refugees have been in since 1959 and the implications this has for other protracted refugee situations as well as for the relationship between the nation-state and discourses on insecurity. The metaphor of "home" is discussed as it relates to yet exceeds our understandings of security, risk, and the nation-state but is situated within an engagement with "a politics of emotion" as the driving force behind making live for protracted displaced populations.

While physical violence is not exercised against Tibetan refugees residing in Nepal and thus they mostly do not fear for their lives when it comes to their interactions with the Nepali government, they nonetheless occupy a state of legal limbo predicated on their inability to advocate for themselves in any meaningful way and on their inability to leave Nepal and resettle in another country. Yet Tibetans in Nepal are spatially and temporally excluded, making their uprootedness doubly nefarious, ironically underwriting a situation in which nationalist discourses within liberal democratic communities undermine liberal freedoms, the right to asylum being one of them.[23] This is supplemented by the suppression of political activities on the territory of Nepal, Kathmandu especially, following the 2008 protests in Lhasa, which spread worldwide[24] and "coincided with the anniversary of a failed 1959 Tibetan uprising against Chinese rule that forced the Dalai Lama to flee to India."[25] The Tibetan refugees' status with the UNHCR is somewhat tentative as well, as "newer" refugees are now defined as persons "of concern to the High Commissioner" and not considered bona fide refugees. In other words and how a member of the TGiE put it to me, because the number of Tibetans arriving in Nepal post-1989 has slowed down to a trickle (around 100 in 2017),[26] they are not seen as legitimate refugee exodus but as individual arrivals leaving Tibet for personal reasons and thus denied refugee status.

THE TIBETAN GOVERNMENT IN EXILE: "REHEARSING THE STATE"[27]

A word on the TGiE would help position the refugees and their legal status better vis-à-vis a discussion on national sovereignty. The TGiE is replete with all the necessary departments and bureaucratic divisions one would expect from an official government. Its main job is stated as being the support and preservation of Tibetan education, health care, language, culture, and traditions for Tibetan diaspora by forging strong ties among Tibetan communities in exile and providing political leadership for the purpose of claims to freedom.[28] Even though the TGiE does not enjoy international recognition, most of its operating budget, in addition to the tax on Tibetans living in exile known as the "green book," comes from foreign donors and is used to help fund the operations of the TGiE offices in India, Nepal, Bhutan, and around the world.[29]

Even though His Holiness the Dalai Lama was the first official leader of the TGiE, over time and under his initiative this changed, culminating in 2011 with the democratic election of Lobsang Sangay as the current president (Sikyong) of the TGiE, divesting His Holiness of any political role in the direction of the TGiE. As one TGiE official in Nepal put it to me, "Even though we are going to be sad when His Holiness passes, we have hope and faith in the TGiE which His Holiness has set up precisely to ensure stability and lasting leadership for the Tibetan people after his death."[30] Thus, despite the fact that Tibetans are a stateless people without an officially recognized government representing them, and even though they have lived in exile for close to sixty years now, they are represented and taken care of by the TGiE. This, alongside the lack of any meaningful political engagement by the Nepali government and lack of territorial claims that conventionally legitimize claims to legitimate authority by governments around the world, has crafted the TGiE as a fairly unique nonstate state.

Their dictates for their "citizens" are also unique, with special attention paid to children attending Tibetan schools, Tibetans speaking the Tibetan language, wearing Tibetan clothes, working in Tibetan handicraft centers, or in various capacities for the TGiE. While it is true that the current TGiE represents the continuation of pre-1959 Tibetan governmental structures, "the prior legal recognition of the state of Tibet is disputed and it certainly no longer exists in any meaningful sense," which makes the TGiE "neither a state *nor* a government."[31] Yet as McConnell has argued, this does not preclude it from acting like a government legitimizing its claim to being state-like but also enacting what a future Tibetan state would and could look like.[32] This parallel reality reveals again how the nation-state has always been under pressure to represent itself "as a unified political entity," yet one that "simply does not exist."[33] In light of this, the TGiE presents the possibility for envisioning an alternative mode of belonging without belonging both to protracted refugee populations and to individuals who live on the margins of society in their own states.

While the TGiE exhibits the political, economic, social, and cultural markers of a state, it is not one, which makes it . . . a quasi-state? An anti-state? An after-state? The territorial recognition that Tibetans desire is contingent on retaining control of some part of Tibet, yet in exile their narratives of belonging outlive pure territorial claims and exist within song, cultural practices, homage to their leader and elders, and a repeated enactment of examples of communal belonging, especially in settlements. This kind of committed transcendence of straightforward sovereign claims to identity formation epitomize a certain kind of resilience and imaginative positioning vis-à-vis structures of power that, while not necessarily bringing Tibetan refugees respite from their exile, does provide them with a continued strong sense of identity even in exile, manifested through practices of everyday Tibet making, as in carpet weaving, traditional clothing days, holiday celebrations, food preparation, and language retention. Are these practices an example of overcoming exile, adapting to exile, or normalizing exile? To what ends are instances of everyday Tibet making examples of a cultural reclaiming of Tibet that defy, if not completely erase, the sense of insecurity that accompanies every exile, especially ones that have been displaced for close to sixty years now? And how are such practices, shared by a community in exile, symptomatic of singular acts of witness to displacement?

Despite numerous critiques of the state and debates as to what it actually constitutes, it is not going away and remains the single more prevailing organizing structure of international relations today and thus cannot be ignored.[34] This in itself raises the first challenge. On the one hand, using the nation-state as the reference point for discussing "the citizen" endorses a narrative placing the former at the helm of one's identity formation, precluding any discussion of cosmopolitanism or difference.[35] In that sense, understanding that being "American," for example, is directly tied to one's belonging to a specific set of historical, social, political, and economic traditions underwriting the American nation-state that in turn inform one's sense of what "being American" means and represents. To deny that fact would be foolish and, not in the least, irresponsible, as millions of people face the barrel of the gun of state armies and dictates in a very real state of "insecurity" born out of nation-state mandates.

Here, the Tibetan refugee population as a whole and their mutually constitutive relationship to the TGiE both underwrite traditional discussions of state power, the TGiE sitting in for a state, while also challenging the straightforward territory-nation-citizen trinity as the TGiE is not an actual state. Having been displaced since 1959, Tibetans in exile lack a specific national geographic marker to hang their claims of statehood on. It does not matter that their displacement is the result of what they perceive to be wrongful, illegal, and exploitative territorial policies on the part of an occupying government; the lack of clear international condemnation of the Chinese coupled with the close to sixty-year exile acts as an ironic legitimizer of the status quo of displacement. The example of the TGiE and Tibetans in exile also problematizes straightforward nation-state claims to lawful territorial command not necessarily because of an ongoing and eventually successful, as in the Bhutanese case, "assimilation" elsewhere, but because of ongoing instances of identity-formation-in-exile and the meaning this has come to hold for what being a Tibetan means.

WITNESSING AND WAITING, WAITING TO WITNESS

The underlying thesis informing this chapter is as follows: the protracted reality of displacement of the Tibetan refugees represents an instance of insecurity defined by temporal displacement, best understood through the idiom of "waiting." Even though they don't face life-threatening physical deprivations, their existence as stateless after an almost sixty-year exile raises two challenges for discussing them alongside other stateless people or discussing statelessness as such, making the discussion in this chapter a timely one. Taking a step away from the nation-state, whether the Nepali one or the TGiE, the Tibetan case is fundamentally one about waiting. I had been struck by this realization during my short field work in 2017 with Syrian and Afghani refugees in Bulgaria who were, for many months at a time, waiting to hear back from the Bulgarian Refugee Agency regarding a decision on their status. The uncertainty they felt was, at its core, informed by not having legal permission to work, not knowing what to do while waiting, and not having any control over their situation or the decisions of the authorities. They could legally neither start a permanent job nor purchase property, open a bank account, drive a car, nor travel freely inside or outside Bulgaria because their status as asylum seekers granted them only the right of temporary stay.[36] The Tibetan case, like with asylum seekers waiting for determination of their status, is exacerbated by a condition of temporal displacement that stands in a powerful way for exile as such. As I learned, this temporal displacement punctuates not only the linear narrative of the nation-state but also the ability of refugees and other displaced populations to provide a satisfactory response to questions about "coping" as "doing."

I had gone to Nepal with the hopes of finding out what Tibetans were "doing" in exile in order to preserve their Tibetan identity. The fact of their prolonged displacement certainly posed a challenge for a traditional interpretation of culture-maintaining practices, and my expectation was that they would, in fact, be struggling to preserve their authentic relationship to being Tibetan. I was also interested in information relating to practices of being Tibetan that could point to singular acts of resistance to their imposed exile. This would have made for an amazing chapter, indeed. However, to my stock questions I received stock answers. The truth I found was in the mundane practices of everyday life, in the repetitive chores of life itself.

***

How do you spend your days?

"Prayer."

"Prostrations."

"Prepare lesson plans."

"Washing, cleaning, cooking, shopping."

***

Like with the TGiE as the personification of Tibet, so too were people's lives informed by what Tibetan culture and traditions had been prior to 1959, but that too was changing. Not only did the refugees perform[37] their Tibetan identity for each other every day but the TGiE, bearing resemblance to any acting government, enacted "everyday techniques of government—naming, reporting, taxing, authorizing."[38]

As has been discussed by McConnell and was confirmed by my interactions with Tibetan refugees in Nepal, the Tibetan exile community exists between two poles: on the one hand, their hope of returning to Tibet and, on the other, their daily work of "carving out a Tibetan way of life in exile,"[39] positioning them on a continuum of insecurity—informed on the one hand by temporal and territorial displacement and, on the other, by the practices that inform the symbolic "inauguration" of a quasi-Tibetan state among Tibetans in exile. The process of waiting for a resolution to their situation adds a layer of complexity not only to discussions of security in the context of the nation-state but also to security that, in the Tibetan case, "both fits and fails to fit notions of state space and the refugee camp."[40]

***

There is a lot of backlash against refugees around the world.
How do you explain that?

"Let's talk about lack of civility. Refugee word has become very ugly. It is because of suspicion which creates fear and disliking. We lack hospitality."

***

In this way, the Tibetan refugees challenge, rather than underwrite, Giorgio Agamben's contention that the camp "is the new biopolitical *nomos* of the planet"[41] in which the production of bare life legitimizes sovereign power and its ability to normalize a "state of exception."[42] In the context of the Tibetan refugee situation, the camp is a reminder of their displacement but it is also imbued with meaning given it by the refugees themselves, a meaning that makes it more "home-like" than unhomely. In other words, I transition from a presentation of the "camp" as a demarcated, already decided space and toward one of the camp as unsettling the category of home that, in the Tibetan case, carries with itself important consequences. This is reminiscent of Salter's discussion of the border as a "suture," in which the latter "better represents the messaging, transition, implication between the inside and the outside"[43] as an active, dynamic, interactive, unsettled space.

But what are Tibetan refugees waiting for and how do their everyday practices of living in exile bear witness not only to the idea of a static state as a problematic conception of sovereign power but also to the various social and emerging mechanisms of "making live" that exceed the state? Fiona McConnell in her truly excellent *Rehearsing the State*, after which this subsection is named, argues that in the case of the TGiE, waiting can be seen in one of two ways: either as an emotionally charging and morally detaching state or as an opportunity to engage in the productive work of "experimentation, preparation and reflection."[44]

***

Do you see yourself as a stateless person?

"I don't know."

"She is feeling sad."

"Yes, yes."

***

***

You don't have a country, you don't have a soil . . . to step on.
If you don't have that, what then?

"As long as our High Lamas are alive, Tibetan culture is not in danger."

"Our High Lamas are coming for reincarnation and they will hold the religion and culture."

***

Because Tibetans technically do not belong to any state, they "belong enough" to China in the eyes of the Chinese authorities to make their relationship to the world one of Chinese priority, enough for Nepal to be welcomed on its territory and tolerated for decades and "enough" to their commitment to a Tibetan solution so that their fate cannot be decided neatly according to either of the three durable solutions outlined in the Convention. In that sense, they cannot return to Tibet even if they wanted to as this would entail asking for travel permission from the Chinese embassy, recusing themselves of any political involvement on behalf of their own claims to freedom and agreeing to forfeit the small amount of freedom they do have in exile.

***

If China invites you to go back tomorrow, would you return?

"As of now, even if we went back, there is not much benefit."

"In the present context there is more disadvantage to go back."

"It is better we should not go without certain rights and freedoms, like in a prison."

***

Also, they cannot be fully and legally assimilated in Nepal, though they have resided there for decades, both because Nepal has not signed the Convention and because the Nepali government has no short- or long-term intention of granting them legitimate status through naturalization. Last, they cannot be resettled in a third country as this would be admitting that China's past and current policies have exemplified persecution. In that sense, Tibetan refugees in Nepal live under the full legal protection neither of the Nepali government, nor the Chinese authorities, nor UNHCR, with the TGiE standing in for the sole guarantor of their rights.

In this context, the Tibetan situation adds another layer to the discussion of risk and insecurity informing this project, a dimension beyond the admission that "risk is another word for power and domination,"[45] even *if* we agree that "nobody is not a risk" and that "everyone poses more or less of a risk for everyone else."[46] The Tibetan population does something more than that: because they are born a(t) risk, theirs is an experience of insecurity informed by a fundamental lack of knowledge of what "not being at risk" means, in which sovereign-guaranteed security almost fades into the background as it has never been available. They represent the exception to the exception, with liberal nation-states taking away and giving back rights to subjects in a fine-tuned game of measured oppression.

Much like the example of children's "war bodies" in chapter 6, whose deaths are discussed in the context of undoing the state, the Tibetan refugees stand in for the state itself as displaced, out of sorts, and refused. I am not saying they do not wish to be full-fledged citizens of a country that cares for them, but they do bear witness to an impotence and a crippling inability on the part of states to care that is characteristic of modern rights givers. So what, some may ask? The so-called undone nation-state goes strong in its mission of denying rights, status, and protection to millions of people. The Tibetan refugees may be displaced, but they bear witness to something else: to a form of solidarity informed by the stateless or the displaced person and helping to reorient the politics of the state as such.[47] To where? Like with waiting and the ways in which it informs one's actual being somewhere, so it is that which stands outside the state that most clearly defines what the state is that casts the most clarity on the true nature of the institution that not only governs but also refuses to govern and as such should be understood as "being produced through dynamic social processes and thus is constantly in emergence."[48]

***

Is Tibet a country or a culture?

"A country."

"I don't know."

***

As this is a chapter informed by personal testimonies, it is important to share an observation on the role of discourse. In the Tibetan case, it is clear that we are dealing with a nation-state that does not speak the same language as the "displaced human" and thus they cannot understand each other, cannot communicate. This may seem obvious to many, but I argue this is an important point that has not really sunk in because if it had, we would pause before marching on constructing solutions not only to displacement but also to terrorism, insecurity, and underdevelopment the way we have always done that: sloppily. Doesn't the fundamental and basic definition we use to recognize and describe refugees and other displaced persons begin with the assumption that what refugees want to recover can be provided by "the state"? This lack of a common language, of a shared meaning is at the core of the impasse that Tibetan refugees find themselves in and is the main reason why holding on to the Tibetan language is so important to them, so central to who they are, and so vital to their survival. This sentiment was reiterated to me time and again in no uncertain terms. Exceeding the state's mandates, one's native language outlives the exception without at the same time normalizing it.

The Tibetan refugees bear witness to a state of belonging nowhere, yet to not being homeless; to a state of not having a "country," yet to a home. How is this possible? In their long-term displacement, they represent an example of "unbounded citizenship"[49] in which one can only engage with them not only through the metaphors of lack and exclusion but also through the larger idioms of community and belonging. Fully aware of the pervasive sense of insecurity that Tibetan refugees in Nepal and elsewhere feel daily, not only are *they* made legible and visible through the lens of nation-state acts of discrimination, as explained earlier, but they also make visible and legible the nation-state's own insecurities, nowhere more evident than when inscribed on the bodies and lives of human beings defined only by exile. From within the holes of sovereign power, through the tears of its tattered cloak, Tibetan refugees imagine the home outside the confines of sovereign security and with relation to language and kin.

***

When were you happiest?

"Last Friday, I went with my friend outside. We had lunch, we were roaming around, free—that was great. I was in my mind free because we were chit chatting girl talk, gossip. We cannot say our lives are long, so we have to be happy whenever we can."

"When I gave birth to my sons and my daughter."

"Yes, I am happy certain percentages. I am disappointed certain percentage because I am a human being, not a robot—neither am I divine. I am mixed. My kids make me happy."

***

Yet they are not wholly and totally beholden to the powers responsible for their condition but also exceed them and not only because of the support and existence of the TGiE. Would it be too much of a reach to suggest that this represents an example of what Antje Ellerman calls "a reverse state of exception,"[50] in which lack of documentation not only affords migrants a certain degree of freedom, a direct result of not being on the official registers and radar of state power as governmentality, but also allows them to exceed "the boundary making and marking"[51] practiced by nation-states? Paradoxically, the possession of a legal identity, so central to rights claims, "constitutes a liability to the illegal migrant,"[52] whereas "the fewer claims migrants can make against the state, the more constrained the power of the liberal state to secure the obedience of the individual."[53] Only in the case of the Tibetan refugees, their persistent lack of documentation is something they lament, seeing it as an imposed state of "dependency" and a "limitation" on their ability to conduct their lives.

***

When was one time you felt humiliated?

"When I travel on the bus and they ask for my papers."

"'I am a refugee'—like someone has slapped you in your face. Using this word again and again you become used to it."

***

In other words, while the protection and provisions the TGiE provides for the refugees are important in their efforts to legitimize claims to being of the Tibetan race, they are met with very real restrictions when it comes to Tibetan residency in Nepal. Their lack of documentation is not the result of their own purposeful destruction of said documents; neither do they interpret it necessarily as a liberating circumstance. Yet their lack of documentation, though a sign of their individual insecurity, also underscores a larger and more pervasive insecurity lurking behind the mask of sovereign power itself. It is a sign, as addressed in McConnell's work,[54] of states engaging in the performance of discursive and symbolic practices of statehood, both at the governmental (machinery of institutions responsible for controlling, surveying, and managing populations) and the sovereign (state as power to punish) level.[55] In the context of this chapter, the Tibetan population in exile exceeds a discussion that reads it simply as a consequence of Chinese aggression, or Chinese rightful claim to territory (whichever the case may be), and positions it more as a symptom of state practices of aggression, exclusion, and violence in general. In this sense, the personal narratives that took me to the Tibetan settlements in Nepal are the same narratives that make it impossible to recognize that "these are narratives that thread the relationships between the bodies and lives of victims through the body and life of the state."[56]

On this journey, I have greatly benefitted from the work of many before me, but one deserves a special mention: James Ferguson and his *The Anti-Politics Machine.* Not only is he critical of economic development assumptions, but he is critical of the Western "development machine" manifesting itself as an anti-politics machine, in which ongoing relations of power are subsumed under the rubric of a benevolent governmental apparatus. His close and meticulous engagement with development's assumptions, goals, mechanisms, and failures pegs against each other several competing discourses. He is a master at showing the workings of power in marginal locations, at the extremities, where it intersects with regional and local practices. He shows that the Thaba-Tseka development project in Lesotho failed not by focusing exclusively on the failures but by unveiling a system of relations beyond state power that, in the end and in the context of Lesotho, upended the development machine itself. Thus, he uncovers a world in which relations of power, meanings, and positions of dominance are constantly challenged, renegotiated, reintroduced, and renegotiated. In the process, he uncovers the real winners of development projects to be the nation-state.

> In this perspective, the "development" apparatus in Lesotho is not a machine for eliminating poverty that is incidentally involved with the state bureaucracy; it is a machine for reinforcing and expanding the exercise of bureaucratic state power, which incidentally takes "poverty" as its point of entry—launching an intervention that may have no effect on the poverty but does in fact have other concrete effects. Such a result may be no part of the planners' intentions—indeed, it almost never is—but resultant systems have an intelligibility of their own.[57]

I argue that the Tibetan example offers a similar opportunity for critique and novel research in which even though the beneficiaries of the international refugee regime are state machines, the latter are problematized in a parallel game of power effects and in which the refugees negotiate their precarious status while rising above it through the "doing" of everyday life. Here, the very status of the nation-state as a giver of rights cannot be upheld unproblematically without also realizing that what states do, above all else, is take rights away. Ferguson's unintended consequences, brought on by the interplay between the development apparatus and local practices, knowledge, and imperatives of the Basotho, are mirrored by the Tibetan example—stateless for their entire lives, refugees stand to offer the most effective, poignant, and timely pronouncement on the nature of state power. The unspoken testimonies stood out the most because they portrayed best my informants as more than "one-dimensional subjects."[58] Their stories bear witness to the obscenity, not totality, of politics—a politics demanding that if I choose to commit myself to narratives about the state, I do so carefully, for fear I leave out "the person" altogether stranded.

> We may convey to readers illusions of truth with our authorial expertise, but often we are the hideous beings swallowed up by our scholarly clothes, the dancing fools under the fluorescent lights of our paradigms and theories that voraciously consume our thoughts, hammer the soul from our words, and drain our voices of any traces of humanity.[59]

What must it feel like as a Tibetan to be asked what feeling like a Tibetan refugee must feel like?

***

What do you think about being a refugee?

"I don't think about it, I am too busy with work."

"Personally, I am very happy to say I am Tibetan and a refugee because of His Holiness."

***

To be able to explain what being a Tibetan refugee feels like, must it not necessarily also heighten one's sense of insecurity, longing, and loss? When called to reflect on the grief, loss, resilience, and confusion that the Tibetans I spoke with felt, there emerged unspoken experiences "whose coherence is not immediately evident."[60] As witnessed by others, "It is not always clear within the contours of trauma and mourning where one ought to begin."[61] But one must begin and, after beginning, commit to finish. And finish I shall, like this:

***

What is your hope for the future?

"Veterinary school. Then go to the US. I want a simple life."

"Study hard. I have my parents and I am happy about that. I want a good man."

"I don't have special hopes. I am thinking about the present. Other people ask this question, as school—stupid question."

"Tomorrow will be better for my children, tomorrow will be very different."

***

Why do you think young people want to go abroad?

"It's a dream. It's a dream.
They are only a fantasy, warped.
They will be exploited, more exploited than they are here.
That they don't understand."

"It is a survival matter now, you have no choice about that.
I understand.
Choice is there when survival is not at stake."

"But the uniqueness will be lost gradually."

And with that?

"It is a big loss for the whole world.
Where to look for example when one country subjugates
Another country and certain number of people are displaced—
From whom will they learn the lesson?"

***

In an effort to understand the unintended consequences of "facts," does an imposed and permanent lack of documentation comment on anything other than the insecurity inherent in lacking documentation? Does it offer a different, if not necessarily better, take on security and its meaning for people who have the luxury of not thinking about it? Does it open up spaces for discussing and interrogating risk not from the point of view of the exception or the norm but from an altogether different point of view of kinship? As academics and policy makers struggle to explain away terrorism, insecurity, and the reasons behind "rational" states' irrational actions, how does the Tibetan example help make the mutually dependent relationship between "lives that matter" and "lives that don't" less of a card in the hand of nation-states?[62]

A Tibet exists in the everyday experiences of Tibetans living in Nepal, where the work of everyday life unfolds despite the practical, legal, political, and economic challenges. Reminiscent of Liisa Malkki's engagement with camp versus town Burundi Hutu refugees in Tanzania,[63] the Tibetans in Nepal exhibit all the pragmatic know-how needed to survive in a foreign country. They have fake documentation, own businesses even though they are legally not allowed to, travel outside Nepal and even beyond India when technically they cannot do that without documentation, and they do have access to banking services, though not always in their own names. Yet this work of everyday life as I call it is also the work of security as "the wise man," the name I gave to this one particular informant, shared: "Risk we can take any time with a secure mind. Security is insecurity, my dear writer, any time in the night, in the afternoon, we feel very insecure. For 59 long years we have been living in Nepal insecured."[64] This insecurity, for him and others of his generation, came from outside but also from within the ranks of Tibetan youth hungry for stability, a job, access to markets, and a modern existence. As with my discussion of survivors of trafficking in human beings whose experiences also made up for a choppy, indeterminate, nonsensical, and confusing narrative, so did refugees' testimony problematize the linear narratives informing most academic work.

The pervasive need to reclaim their homeland, to know it and walk it again remains strong for the older generation of Tibetans who have recollections of their time in Tibet. The younger generations, however, never having set foot on Tibetan soil and knowing their "homeland" mostly as an aspiration, found it harder to identify with the romantic vision of free Tibet their elders espoused for them. My friend Tashi, who was thirty-five at the time of our interview, shared that even though his parents still believe that Tibet would be liberated one day, he does not. Yet Tashi and others also shared that teaching, which is what he did for a job, was important to them because the Tibetan cause gave them purpose.

***

What is the main way of preserving your culture when you don't have a territory to go back to and when most of you have never been to Tibet?

"It is important to stay in settlements to feel Tibetan."

"Culture, language, food, traditional clothes."

***

Even though it is true that in displacement, refugees are "rendered politically insignificant,"[65] the Tibetan refugees are truly an exception to *that* norm in that they cannot simply be categorized as "bare life." Their engagements with the state and with sovereign power do not constitute desperate measures, nor are they specifically aimed at challenging and usurping the powers of the nation-state per se. While their legal situation in Nepal is truly precarious, they indicated they did not always feel stateless, no matter how many times I pressed the issue. Sure, they exist in a legal limbo, but they are also supported and looked after by the TGiE, so their minimal dependence on the Nepali state actually works to solidify the latter's lack of authority over their well-being.

This, however, is not true of newly arrived (post-1989) Tibetans in Nepal who find themselves categorized, or shall I say uncategorized, in an additional way, according to the willingness of host governments and international organizations to honor their claims to rightful asylum based on the calculations and arbitrary decisions of bureaucrats. Are they "displaced non-official-refugees of concern for High Commissioner"? Or is it the fact that they don't belong to any one category that makes them simply Tibetan? In the context of a discussion on insecurity and risk vis-à-vis the modern state, displaced Tibetans clearly identify themselves as such, even as their situation is considered "a political case" and thus unresolvable.

It is important to note that even though no law denies them status, no law gives them one; they are not altogether outside legal norms. Interestingly enough, while one TGiE official shared that he was unable to look after the scattered refugees but only after the settlement residents, another shared that in fact it was the scattered refugees who could do pretty much whatever they wanted while the settlement residents had to follow rules and norms. In other words, noninterference by the Nepali government is earned as a condition of Tibetan noninterference with the Nepali state. The status quo that ensues as mutual toleration allows the TGiE to operate in Nepal, which is truly exceptional in this case.

While the Dalai Lama merits special attention with regard to his ability to maintain control over, provide guidance for, and influence Tibetans in exile worldwide, he also represents a case of a sovereign-in-exile who has managed to recreate and envision Tibet for his people beyond the confines of territorial states and enclosed spaces. In this way, I read the Dalai Lama's legacy as one spanning beyond his statute as the spiritual leader of his people, and more as providing the possibility to reenvision and reassert Tibetans' role in the world, whether they are residing in Nepal, India, Bhutan, or Boulder, Colorado.

***

Do Tibetans in India feel more secured?

"It is the same.
. . .
And thinking about children
is also a difficulty,
they were born in exile.

To them the word refugee,
the intensity of the word
is not fully applicable.

To us, elderly people,
the intensity is very great,
we have the fear that we may lose
our old identity, our tradition.

We also do have the fear that
youngsters might do something drastic
which is against tradition,
deviate from the path.

It is also a fear."

***

In other words, in the absence of papers, refugee cards, and recognition in the official, juridico-political sense of the word, Tibetans and their reliance on the Dalai Lama for guidance and support has made visible for them the possibility of a creative nonbelonging—one that honors the host country but that is also largely independent and autonomous from host country attitudes. While the gentleman's agreement between Nepal and the UNHCR is really quite extraordinary, considering the fact that Nepal is not under any obligation to keep it, it is also true that Tibetan refugees' situation cannot simply be limited to their juridico-political status and possession of valid documentation. Along the lines of Ellerman's argument relating to the possibility of resistance by destroying migrant documents,[66] the Tibetan example is one in which the lack of documentation, which is inconvenient, does not fully impede Tibetans' ability to maintain their culture and lead productive lives, though their relationship to state structures are strained. In a sense, they exist in a juridico-political lacuna, a blank space, in which the regular laws of sovereign states sort of apply but not really. And so, exiled from their home country, they exist on the margins of sovereign power and in that everyday existence create spaces for everyday perseverance that, in the absence of meaningful ways of resisting, becomes the next best thing.

CONCLUSION

Tibet thus lives in a performative register of being Tibetan by Tibetans in exile rather than as an actual tangible place, site, and location. Tibetan refugees feel a strong connection to the cultural heritage of their motherland, which is informed by memories of Tibet, images of His Holiness, practice of traditions such as consumption of traditional Tibetan food and wearing of traditional Tibetan clothes. In a conversation with a representative of the Snowlion Foundation, the fund-raising and educational, health, and social development wing of the TGiE, to my question relating to Chinese influence in Tibet, I received the following answer: "Making roads and buildings is not progress. When money speaks, the truth is silent." An insistence on the power of truth, cultural, and moral values was also a frequent response to my questions regarding suffering and resilience.

***

What do you think about the self-immolations in Tibet?

"No one can subjugate the human soul . . . even fear."

***

The second most common answer, after the one relating to the health and blessings of the Dalai Lama, had to do with a belief in doing the right thing, in policies of nonviolence, kindness, and compromise, and an unwavering faith in a life of service to the Tibetan community.

Curiously, the camp manager of one of the Pokhara camps I visited commented on how truly precious Tibetans must be in the eyes of the Chinese for the latter to take so much care, through aid to Nepal, to keep Tibetans quiet and in Nepal. On the other hand, a headmaster of one of the Tibetan high schools offered the following observation: "People should not emphasize kindness and compassion so much. It is ok to express your feelings. Rather, emphasize working hard, success, perseverance, and for students to be confident and not to repress their feelings." He proceeded by turning the political focus away from Nepal, admitting that the Nepali government is totally useless when it comes to resolving the Nepal-Tibet situation, and that unless the international community gets involved, nothing would be accomplished. He concluded that the road to empowerment was through education and personal empowerment so that people could be effective for themselves, regardless of China, Nepal, or India. To him, this personal success was the best possible way to resist exile and build resilience as Tibetans.

Finally, the situation of the Tibetan refugees has been a hot topic for such a long time that there seems to exist a mythology about them and what their stories ought to say. In my case, I expected them to address statelessness, territorial belonging, and issues of political rights, and a few of the employees of the TGiE did. But the majority of the refugees could care less about these issues, either because they had never thought of them or because they were just unwilling to wallow in such unimportant talk. I discovered that their main objective was survival and as they felt a strong connection to their cultural, religious, and spiritual traditions, their sense of self stood divorced and detached from questions of territory as we know it in the West.[67]

***

Do you have a question for me?

"No, I don't."

"I don't think so."

"No, no, no."

"No question, no."

"I think I have told you a lot of things about myself."

***

NOTES

1. "Tibetan Refugees in Nepal," http://www.tibetanreview.net/tibetan-refugees-in-nepal/ (accessed February 1, 2018).
2. Uma Shankar Prasad, "Study of Nepal's Economic Relations with China," *Journal of Development and Administrative Studies* 23 (2015): 23–32.
3. "The Tale of One Tibetan Refugee," http://utsynergyjournal.org/2017/05/06/the-tale-of-one-tibetan-refugee-in-nepal-memories-insecurity-and-home/ (accessed January 27, 2018).
4. "The Tale of One Tibetan Refugee"; "A State within a State: Tibetans in Nepal," https://thediplomat.com/2016/07/a-state-within-a-state-tibetans-in-nepal/ (accessed January 27, 2018).
5. Government of Nepal, Ministry of Finance, Documents—IECCD Publications, http://www.mof.gov.np/en/document/?c=65&t=&y= (accessed February 14, 2018); Prasad, "Study of Nepal's Economic Relations with China," 28.
6. "India's Aid to Nepal Dwindles, China Now among Major Donors," https://economictimes.indiatimes.com/news/politics-and-nation/indias-aid-to-nepal-dwindles-china-now-among-major-donors/articleshow/51984004.cms (accessed February 14, 2018).
7. "Nepal Top Recipient of India Aid with Rs 253 Crore," https://www.hindustantimes.com/india-news/nepal-top-recipient-of-india-aid-with-rs-253-crore/story-0HFcfbgSj9zaGVbuE4gASM.html (accessed August 24, 2018).
8. "India's Aid to Nepal Dwindles, China Now among Major Donors."
9. Prasad, "Study of Nepal's Economic Relations with China," 26.
10. "High-Level Chinese Visit to Nepal Highlights Difficulties for Tibetan Community," https://www.savetibet.org/high-level-chinese-visit-to-nepal-highlights-difficulties-for-tibetan-community/ (accessed February 14, 2018).
11. "China Should Offer More Generous Aid to Nepal amid Sino-Indian Border Dispute," http://www.globaltimes.cn/content/1061315.shtml (accessed February 14, 2018).
12. Prasad, "Study of Nepal's Economic Relations with China," 27.
13. "Nepal to See China's Assistance for Poverty Alleviation in Border Districts," http://www.xinhuanet.com/english/2017-05/30/c_136326330.htm (accessed February 14, 2018).

14. UN General Assembly, *Convention Relating to the Status of Refugees*, July 28, 1951, United Nations, Treaty Series, vol. 189, 137, http://www.refworld.org/docid/3be01b964.html (accessed September 24, 2018).

15. Alexandria J. Innes, *Migration, Citizenship and the Challenge for Security* (London: Palgrave Macmillan, 2015), 95.

16. Fiona McConnell, *Rehearsing the State* (Oxford: Wiley Blackwell, 2016), 22.

17. "World Refugee Day 2017: Nepal Stands with Refugees," https://www.spotlightnepal.com/2017/06/17/world-refugee-day-2017-nepal-stands-refugees/ (accessed January 27, 2018).

18. "History of the Bhutanese Refugee Camps," http://bhutaneserefugees.com/history/ (accessed January 27, 2018).

19. Ibid.

20. "Tibetan Refugees in Nepal."

21. Author interview on May 31, 2017, at Tashi Pelkhiel Tibetan Refugee Settlement, Pokhara, Nepal.

22. Excerpt from a speech by the Dalai Lama that I copied from a Tibetan language textbook at Atisha Primary School in Jawalakhel Tibetan Refugee Settlement, Lalitpur, Kathmandu.

23. Innes, *Migration, Citizenship and the Challenge for Security*, 80.

24. "Violence in Tibet as Monks Clash with the Police," http://www.nytimes.com/2008/03/15/world/asia/15tibet.html (accessed February 10, 2018).

25. Ibid.

26. "The Tale of One Tibetan Refugee."

27. This is a direct reference to Fiona McConnell's *Rehearsing the State*.

28. Central Tibetan Administration, http://tibet.net/ (accessed January 23, 2018).

29. "Exile Tibetan Parliament Passes Budget of 2,434 Million Rupees," https://www.tibetsun.com/news/2017/03/26/exile-tibetan-parliament-passes-budget-of-2434-million-rupees (accessed February 26, 2018); Central Tibetan Administration, Department of Finance, http://tibet.net/department/finance/.

30. Author interview on May 22, 2017, in TGIE office in Kathmandu, Nepal.

31. McConnell, *Rehearsing the State*, 28.

32. Ibid., 28–29.

33. Ibid., 17, 31.

34. Judith Butler and Athena Athanassiou, *Dispossession: The Performance in the Political* (London: Polity, 2013).

35. Jacques Rancière, "Who Is the Subject of the Rights of Man?" *South Atlantic Quarterly* 103, no. 2/3 (2004): 297–310; Innes, *Migration, Citizenship and the Challenge for Security.*

36. Innes, *Migration, Citizenship and the Challenge for Security*, 90–103.

37. Judith Butler, "Performativity, Precarity and Sexual Politics," *AIBR: Revista de Antropología Iberoamericana* 4, no. 3 (2009): i–xiii.

38. McConnell, *Rehearsing the State*, 32.

39. Ibid., 33.

40. Ibid., 63.

41. Giorgio Agamben, *Means Without End* (Minneapolis: University of Minnesota Press, 2010), 44.

42. Giorgio Agamben, *Homo Sacer: Sovereign Power and Bare Life* (Stanford, CA: Stanford University Press, 1998); Giorgio Agamben, *State of Exception* (Chicago: University of Chicago Press, 2005).

43. Michael Salter, "Theory of / : The Suture and Critical Border Studies," *Geopolitics* 17 (2012): 738.

44. McConnell, *Rehearsing the State*, 23.

45. Ulrich Beck, *World at Risk* (Cambridge: Polity Press, 2009), 142.

46. Ibid., 187.

47. Jacques Derrida, *On Cosmopolitanism and Forgiveness* (New York: Routledge, 2001), 4.

48. McConnell, *Rehearsing the State*, 31.

49. Linda Bosniak, *The Citizen and the Alien: Dilemmas of Contemporary Membership* (Princeton, NJ: Princeton University Press, 2008).
50. Antje Ellerman, "Undocumented Migrants and Resistance to the Liberal State," *Politics and Society* 38, no. 3 (2010): 408–29.
51. Elizabeth Dauphinée and Christina Masters, eds., *The Logics of Biopower and the War on Terror* (London: Palgrave Macmillan, 2007), 232.
52. Ellerman, "Undocumented Migrants and Resistance to the Liberal State," 1.
53. Ibid., 2.
54. McConnell, *Rehearsing the State*.
55. Michel Foucault, *Discipline and Punish: The Birth of the Prison* (London: Vintage Books, 1995).
56. Dauphinée and Masters, *The Logics of Biopower and the War on Terror*, ix.
57. James Ferguson, *The Anti-Politics Machine* (Minneapolis: University of Minnesota Press, 1994), 255–56.
58. Dauphinée and Masters, *The Logics of Biopower and the War on Terror*.
59. Roxanne Lynn Doty, "Maladies of Our Souls: Identity and Voice in the Writing of Academic International Relations," *Cambridge Review of International Affairs* 17, no. 2 (2004): 378.
60. Dauphinée and Masters, *The Logics of Biopower and the War on Terror*, xvi.
61. Ibid., xv.
62. Ibid., ix.
63. Liisa Malkki, *Purity and Exile: Violence, Memory, and National Cosmology among Hutu Refugees in Tanzania* (Chicago: University of Chicago Press, 1995).
64. Author interview in Tashi Ling Refugee Settlement in Pokhara, Nepal, on May 31, 2017.
65. Ellerman, "Undocumented Migrants and Resistance to the Liberal State," 2.
66. Ibid.
67. I would like to thank my former student and good friend Stefano Ballesteros for shedding important light on the connections made here between territory, sovereignty, and identity.

Chapter Six

The Child's Dead Body as "War Body"

This is a chapter about death. This is also a chapter about our disciplinary relations to war as it is inscribed on the bodies of children victims of war either as bystanders to active conflict or in the process of migrating because of violent conflict. If "understanding people's experiences with/in war is essential for understanding war,"[1] I argue that children's deaths in war, intended or not, represent a crisis of representation for scholars of international relations (IR) and point to two ongoing puzzles. The first one deals with the fact that war is anything but a reasonable, rational, and civil affair and thus cannot be investigated simply through the lens of IR theoretical commitments to the conclusions of scientific knowledge. Second, it points to the precarious position that children occupy within IR when it comes to conflict, risk, and insecurity. While children (zero to eighteen years of age) are usually legal subjects of the countries of their birth, they are always only quasi-citizens in the fully legal sense of the term. In legal, political, and economic parlance, as minors they are not allowed to vote, run for office, have a bank account, hold a job (up to a certain age), own property, marry (age dependent per country), or drive a car, to name a few. They do, however, have rights specific to them and granted internationally through the United Nations Convention on the Rights of the Child[2] (referred to from here onward as "the Convention"), adopted in 1989 and lacking the signature only of the United States. According to the language of the Convention, discussed in detail later in the chapter, children are considered vulnerable, fragile, dependent, weak, and in need of protection, no matter whether at war or in peacetime, validating the claims of nation-states as well as parents and society to their obligatory and required care and patronage.

The chapter argues that representing children's deaths at war as collateral damage and an unintended consequence of violent conflict just cannot do.

Embracing the finality of the human body, coupled with the very real physical, mental, and psychological flux and indeterminacy of childhood, combined with the monstrosity of war and the insistence on promoting and making excuses for the priorities and goals of nation-states and other perpetrators of violence, is something that warrants engagement. I am here privileging the lifeless bodies of toddlers washed on the shores of Greece and Turkey, the shrapnel-covered body of six-year-old Yemeni boy Fareed Shawky, first "saved" only to succumb to death days later, offer no easy answers to the relationship between IR and war, because these do not fit neatly within the IR register.

I continue to argue that engaging risk through one's material, corporeal essence offers the most authentic way for human beings to learn, experience, navigate, and problematize politics. Why the need to problematize? Because it's problematic already, and because the inherently singular nature of the body, which we all share, helps us make sense of the world. In this sense, "violence is specifically destructive of the ability of the body to engage in sense-making, as it 'undoes our deployment of the bodily' I can in making sense of the surrounding world."[3] Thus, bodies not only live through and navigate the natural processes of birth, growth, and death, but they also possess intimate knowledge of "risk" and its daily performative inscriptions on human flesh. The knowledge contained in children's deaths further complicates discussion of war and warrants its own engagement.

With reference to war deaths in general, the representation of male bodies, a long-standing symbol for "the soldier," during war is well documented, both in terms of performing masculinity and in terms of masculinity's appropriation during battle not as wounded masculinity but as a symbol of national pride and heroism.[4] In this sense, the wounded and dying soldier, though shocking to the eye, does not necessarily figure (when perhaps it should) as a particularly unusual and/or novel picture of horror in contemporary military discourse. Compared to my discussion of trauma and victimhood incurred as a result of trafficking or displacement, death in war epitomizes the ultimate traumatic experience perhaps because of its finality. Does this finality, this fatality make theory obsolete? Much has also been written on the relationship between war and the nation-state, pitting the destructiveness, chaos, and trauma of the former against the constructed determinacy, wholeness, and integrity of the latter.[5] Yet "war unsettles the delicate collective trust in the practices of meaning that constantly produce the imaginary of coherent and stable national identity"[6] and thus uncovers not only the insecurity inherent in the nation-state but also the need for reinterrogating the meaning of this imagined "stable national identity" precisely from the point of view of those living on its fringes. The goal of this chapter is to examine the relationship between death, the child, and the negotiation and renegotiation of national identity through a focus on cases from recent Mediterranean Sea ("the Sea" from here onward) boat crossings and the wars in Syria and Yemen. What I hope to show is the inherent vulnerability, fragility, and indeterminacy of the nation-state narrative itself, coupled with a discussion of the dead children's bodies in question in an effort to suggest an engagement not with bodies as corpses but with corpses as bodies. How do the images of drowned toddlers washed on the shores of the Sea, for example, arrest the unproblematic narrative of IR as the study of the behavior and nature of nation-states, as the study of normalized, accidental, or extraordinary violence? The lifeless bodies of children serve as a reminder both of the spectacular nature of war/conflict and the realization that "people live in wars, with wars, and war lives with them long after it ends."[7] The everyday realities of the boys, girls, and their families fleeing conflict in Syria, Afghanistan, Yemen, or Myanmar, to name a few, exist in a kind of perverse intimacy with death already, exemplified and materialized for us in the images of dead children and having already been present in these children's lives for a long time.

Achille Mbembé is highly informative when he asks, "But under what practical conditions is the right to kill, to allow to live, or to expose to death exercised? Who is the subject of this right? What does the implementation of such a right tell us about the person who is thus put to death and about the relation of enmity that sets that person against his or her murderer?"[8] Life and death are discussed as both inscribed and excluded from the order of sovereign power, understood under the banner of "necropolitics" as an alternative to "biopolitics"—the former standing in for the relationship between sovereignty and the power over life and death, in which "sovereignty means the capacity to define who matters and who does not, who is *disposable* and who is not."[9]

Building on this idea, I borrow here from Christine Sylvester's writings on the relationship between IR and the everyday experiencing of war, experiences that "come in prosaic, profound, sickening, excruciating, and exhilarating ways, to all kinds of people living inside and outside war zones."[10] Constructed as unnecessary, redundant,[11] and dangerous, refugees and other migrants figure on the public horizon as a threat.[12] In the European context, "elites promote certain kinds of freedom among the governed," which is predicated on excluding "the poor and the 'ignorant majority'" from their ability to make "demands and express views that could lead to chaos, undermining the pursuit of national (and elite) economic efficiency."[13] Reduced to "bare humanity," refugees and migrants require attention and care, while at the same time bringing not only limited resources but also limited information related to their past, the levels of threat they pose to the host nation, and their intentions for engaging the latter. These "unknown unknowns" make their plight unresolvable and their difference and needs a burden. They are constructed as "risky bodies," which necessitate appropriate responses on the part of the nation-state in mitigating the threat they pose and neutralizing their political charge vis-à-vis the native citizens. They are called, among other things, "exceptional" and carry within them the charge of the exception as conceptualized by Giorgio Agamben.[14] They exist outside the law and can be disposed of, deported, and/or even killed without recourse to retribution and without revenge. But that is not all—if Agamben provides the foundation for thinking about their relationship to sovereign power, he also provides an impetus for outliving that foundation. While I have spent much of this book engaging with Agamben's work in a critical way, I bear witness to the fact that his writings contain within themselves the restlessness that informs their ongoing pertinence and relevance.

It is in this sense that Judith Butler's discussion of "ungrievable lives"[15] helps move us beyond the politics of the exceptional and toward a politics of meaningful engagement with people's plight and suffering, with their mourning and loss—all constitutive parts of the makeup of international relations as well. "If 200,000 Iraqi children were killed during the Gulf War and its aftermath, do we have an image, a frame for any of those lives, singly or collectively? Is there a story we might find about those deaths in the media? Are there names attached to those children?"[16] If we don't have an image or someone to bear witness and mourn these children, what becomes of their memory and how are we to address and understand their bodies prior to and in death? Trevor Noah, the current host of *The Daily Show*, provides an interesting observation, also an ironic one, regarding the status of black South Africans during Apartheid, where what contributed to their oblivion was the fact that they were not counted in the same ways that, for example, the Nazi war machine counted the Jews. In a sense, to be counted is to be included and recognized.[17] The facelessness of 200,000 Iraqi children that Butler evokes takes us beyond this, to a place where the modern machinery of state-sponsored death reifies the existence of those it obliterates precisely by counting them, almost like the state claims sovereignty over death as well.

Thinking of refugees and the displaced as "war bodies" helps to reconceptualize their experience with danger long before they come in contact with the camera, long before one of their kind dies tragically, and with an eye to how everyday stories of suffering and displacement lie at the heart of understanding not only insecurity, risk, and trauma but also politics itself. "War bodies are not all lifeless and mutilated; some are warriors, injured, crippled, raped, held hostage, spying and spied upon, grieving and even celebrating."[18] "War bodies" perhaps do the best job at exposing and, in doing so, undoing the myth of the nation-state as protector in showcasing how the latter either directly threatens the human body or is unable to protect it. Where do children's bodies fit in all this? How do children who find themselves the victims of violence inform about state (in)security?

I won't hide that I have been baffled, terrified, saddened, and gutted, as a human being and a parent, by images of dead and injured children around the world. Their lifeless bodies, as I put my own children to sleep at night, have been hard to rationalize, engage with academically, and explain away as "collateral damage." They have haunted me and continue to do so. Perhaps the crude emotional charge they evoke will be considered "cheesy" and a romanticization, but I accept the charge—and try not to do them injustice all the same. The shock has been not because children don't die, because they do, around the world every day of hunger, disease, violence, or accidents. My shock doesn't come from a belief that children's lives are more precious than other lives and are therefore exceptional. Their civilian and minor status does not presuppose or imply exceptionality, nor is their vulnerability in the face of violence unusual.[19] As we cannot really "pick and choose our involvement with the state," and as I am weary, alongside Athena Athanassiou, of all political claims being reduced to "claims of recognition," the children's deaths bring up a tension affirming the importance of "the struggle with the norm, the struggle implicated in that which it seeks to contest."[20]

Before I proceed, a word on the extent of the violence toward children referred to earlier. The statistics for 2017 alone, the year that has ended most recently at the time of writing, tell the story:

> In Afghanistan, almost 700 children were killed in the first nine months of the year. In the Central African Republic, children were killed, raped, abducted and recruited by armed groups. Boko Haram, a jihadist organisation, forced at least 135 children to act as suicide bombers, nearly five times the number in 2016. Violence drove 850,000 children from their homes, while more than 200 health centres and 400 schools were attacked in the Democratic Republic of Congo. In South Sudan, more than 19,000 children have been recruited into armed forces, and over 2,300 children have been killed or injured since December 2013. In Somalia, nearly 1,800 children were recruited to fight in the first 10 months of the year. Nearly 1,000 days of fighting in Yemen left at least 5,000 children dead or injured, and more than 11 million children needed humanitarian assistance. Around 220,000 children lived under constant threat of mines and under explosive remnants of war in eastern Ukraine.[21]

What do these statistics show beyond the numbers? Why the need to list, enumerate, and shock? Shocked, how do we comport ourselves in the face of such violence? Do we normalize it? We do. Do we read more and watch more and learn more and become numb? We do. Do we ignore it? We do that too.

The numbers don't tell much. Nine hundred children killed in Afghanistan in the first nine months of the year makes for one hundred killed per month, or around three per day. Afghanistan is a war-torn country and violence there is rampant, so the children's deaths, though shocking, should not be entirely surprising. Compared to the 923 children who died as a result of the war in Afghanistan in 2016, 2017 has shown an upward trend, especially if we add to this the close to 2,600 children injured in 2016, a rise from 2015 still.[22] Child soldiers, human shields, suicide bombers, child brides, servants, bystanders—the roles that children play in war today have truly acquired a profundity unmatched by any previous period in history. In this role, children's dead bodies ought to be read as "war bodies" that convey meaning that most scholars and, for that matter, most ordinary people are not used to or willing to discuss. Yet children's deaths easily enter the register of what is unspeakable, which, like Giorgio Agamben's *Muselmänner*,[23] relegates them to an ironic silence and a ban, a forgetting.

Wishing to break this silence, the chapter unfolds in the following ways. First, I briefly discuss accounts of soldier bodies at war, starting with World War I and finishing with accounts of veterans of the Iraq and Afghanistan wars. Then I explore past and ongoing instances of committing violence to the bodies of children more specifically. The next part of the discussion engages with the bodies of Aylan Kurdi, Fareed Shawky, and Omran Daqneesh in an attempt to uncover what their deaths expose in the context of IR's preoccupation with providing, disseminating, and ensuring security. The chapter concludes by extending an invitation to IR to acknowledge war narratives as sites of death, as unsettling and troubling, specifically as the latter "construct the international in various ways and are powerful courses or symbols and meanings."[24] In the end, to honor the child's lifeless body, an engagement with children victims of war must be supplemented by an engagement with what the dead bodies symbolize: the temporality, fragility, and contingency of *the nation-state* itself.

THE BODY AT WAR

> War, after all, is as much a means of achieving sovereignty as a way of exercising the right to kill.
> —Achille Mbembé, "Necropolitics"

> How does one, might one, should one look at images of the mutilated, diseased, disfigured or dead body?
> —Debra Lennard, "Censored Flesh"

"The main purpose and outcome of war is injuring."[25] The fact that soldiers' jobs are first and foremost to kill and destroy other human beings ought to give us pause when valorizing and constructing narratives of heroism. Yet "one can read many pages of a historic or strategic account of a particular military campaign . . . without encountering the acknowledgement that the purpose of the event described is to alter (to burn, to blast, to shell, to cut) human tissue."[26] Debra Lennard's excellent analysis of visual representations of mutilated and dead bodies in World War I offers a glimpse into the formation of a crisis of representation, a crisis that is now older than one hundred years. Her engagement with images of shocked, dying, mutilated, injured, and abandoned soldiers uncovers the ways in which the war body during World War I was shrouded in mystery, censored, and sterilized prior to its appearing for the general public. "The war-wounded body was considered 'dreadful' and 'dangerous to the eye,' as viewing it caused horror and subsequent abjection,"[27] facing both the artist and the spectator with the challenge of either repressing or consuming this great horror. Thus, "*not looking* defines the public response to facial disfigurements in Britain"[28] during and right after World War I.

On the other hand, World War I also instantiated discourses tying war to masculinity, in which "masculinity underpinned imperialism and militarism, fundamentally shaping the experience of modern was as social, embodied and psychological experiences."[29] The representation of World War I as a modern, technological war contrasted with soldiers' traumatic physical and psychological experiences of the war, in which "shellshock as a result of modern war was situated within both class and gender frameworks, as a 'crisis' of masculinity,"[30] while conscientious objectors were constructed as "cowards" and "effeminate" for their refusal to stand behind the war effort and their fundamental opposition to ideas of "the citizen soldier's masculinity."[31] The "soldier's body" thus bears resemblance but is not the same as "the war body" in which the former is characterized by wholeness, valor, and courage and the latter is often wounded, destroyed, and maimed. Jenny Edkins merits quoting here at some length:

> In the end, what is at stake is not that the figure of the disfigured veteran reminds us of war, which we would rather forget, or that the face transplant recipient might look like the donor. . . . The disfigured and the faceless remind us of our own animality, vulnerability, and interconnectedness. . . . The disfigured do not reflect that misapprehension, but rather remind us of our fragility of being. It is not the other that appears monstrous, initially, but we ourselves. . . . We cannot recognize the disfigured.[32]

Additionally, Suzannah Biernoff argues that extreme suffering and the violation to the face, the touchstone of humanity, undercut fantasies about heroic warfare and the capacity of military masculinity to withstand modern technological warfare.[33] The modern interventions allued to in Edkins's quote don't make the situation much better either. In her *Portraits of Violence: War and the Aesthetics of Disfigurement*, she builds on Edkins's argument by stating that "the military body is more than flesh and blood: it is a symbolic site invested with political as well as personal meaning. Valour, heroism, patriotism, courage: these concepts assume visible form, and do their cultural work, when they are personified and embodied."[34] It is this making visible that children's "war bodies" epitomize, especially as they reveal themselves to us without a preset notion of what they mean and signify individually as well as more universally in the context of war.

In an account of German prisoners of war in the United States during World War II, Matthias Reiss traces and engages perceptions of the German soldier's body in the imaginary of everyday Americans. His engagement with German bodies is central to his understanding of the experience of captivity. The Germans, perceived as monsters and superhuman giants at the same time, were actually seen "just like any human being."[35] A reporter's experience at the POW camp Gruber is indicative of the symbolic representation and understanding of German soldiers as American "boys."

> Uniformly neat, excessively polite, splendidly disciplined, these young men are—frankly hard to dislike. Whatever poisonous ideologies may still be boiling underneath their sun-bleached hair to all outward appearances they are like any American kids from Tampa to Tacoma.[36]

In other words, the trained, fit, expert, and disciplined German POW not only aligned with the vision of the average American young man, but he also exemplified an epitome of masculinity, the body's appearance central to this narrative. Physical features were constructed as more important than ideological differences, inviting the observer to conclude that because these young men were "uniformly neat, excessively polite, splendidly disciplined," they were also of "good character."

This example portrays the male body at war as a normal site of violence, sacrificed in the name of the nation even when mutilated, executed, and violated. In important ways, the male body's sacrifice during war embodies the nation-state narrative of selflessly sacrificing oneself for the betterment of the country. In this sense, Moustafa Bayoumi strikes a chord when he writes: "We have expectations for our war stories. They should be full of heroism and sacrifice, along with recognition that combat is horrific. Our love for these stories is insatiable, as if we believe the nobility of the soldier is a way to convince us of our own national decency."[37] This is a remnant from classical Greek education, in which the body was seen as "a crucial element in the 'typical' procedures of the 'political' life, as having a clear political function and utility, since the body was simultaneously an agent of political power and subject to the discipline of the *polis*—the body was committed to the decisions of the *polis*."[38] The reification of the dead body and the ensuing claim to sovereignty over death by the state are here historically situated. While the goal of this chapter is not to analyze who benefits and who does not from such power, it is nonetheless useful to remember that bodies at war represent an ideological extension of the nation-state and thus perpetuate a view of the warrior as masculine, stable, functional, noble, and whole, all qualities inscribed on the state in return. In this sense, how do the mutilated/dead bodies of soldiers, civilians, and children rewrite this relationship? The unambiguous way of presenting the soldier's body at war as heroic and courageous tells only half the story.

Michel Foucault has discussed the body as a site of political power, both subjected and subject to the disciplining mechanism of sovereign (pure force) and governmental (mechanisms and apparatuses of control) power, mobilizing, and immobilizing the body proper, inscribing itself on its surface and beneath. Used as a trope for a stable, whole, and permanent national identity, the body of the male soldier has served to produce collective fantasies of communities, shared national identities, and ideas of belongingness[39] as the latter reach "into the very grain of individuals, touching their bodies and inserting itself into their actions and attitudes, their discourses, learning processes and everyday lives."[40]

This is easy to see in image after image of American diseased soldiers and their bodies shrouded in the American flag, replacing a focus on their dead bodies, in a truly Periclean fashion, with a focus on civic duty and sacrifice, invoking their bodies "as metonyms of the nation-state."[41] Thus, the originary purpose, the strategic *telos* of the body of the soldier is again, in life as in death, situated within the nation-state and its ongoing production and display of the wholeness and integrity of the social order. In other words, "the military embraces a model in which the personal body is identical to the body politic,"[42] the latter unified under the banner of one war or another. The managing of injury, conflict, and battlefield wounds evolves into a trope of homogenization, a process indicative of armed forces' transformation of "civilians" into "soldiers." With reference to the Vietnam War, in *Born on the Fourth of July*, Ron Kovic's description of the mechanized induction of young bodies into the Marine Corps further witnesses the production of soldiers as "interchangeable parts of a military machine."[43] In a sense, war blurs the lines between the acceptable and the taboo, as soldiers are inducted into the bomb-throwing, gun-firing, bad-guy-shooting mentality of the unit. The erotic masculinity of the soldier, referring back to my earlier discussion of German POWs in the United States during World War II, is replicated in the bodies of all soldiers: fit, determined, ready for combat, courageous, and invincible.

If human experiences, whether positive or negative, help humans make sense of the world, I suggest that war is perhaps one of the ultimately singular ways of experiencing the world, our vulnerability and our fragility in it. Elaine Scarry makes a similar point in *The Body in Pain* when she claims that the injured bodies as well as injuring bodies during war are at the heart of war as an "act of deconstruction."[44] In death, the war body is separated from itself and "conferred to an ideology or issue or instance of political authority impatient of, or deserted by, benign forces of substantiation."[45] Kevin McSorley's analysis of Scarry's usefulness for constructing a sociology of war suggests that "the out-injured parties in war are no longer able to make the same sense of the world and as such are less able to resist the imposition of a new political reality."[46] Not only is the soldier's body instrumental in the soldier's ability to remake sense of the world, it also becomes instrumental in the ability of nonsoldier bodies and other "war bodies" to make sense both of war as deployed on TV and in newspapers. A perfect example of civilians' difficulty in making sense of the terrors of war is the unbelieving, speechless reaction by the entire American public to the 9/11 terrorist attacks and the ensuing hunt for terrorists culminating not only with beheaded bodies in Iraq and Afghanistan but also with gruesome scenes of torture on U.S. soil.

Relying on the construction of a masculinity at war not only shapes narratives of war and the nation but also influences the representation of veterans' bodies to the public. In this sense, the injured veteran's body not only exists outside the rhetoric of heroic sacrifice but is also materially and corporeally beyond the scope of the spectacle of war, augmented by an ever-expanding field of technological advancements enhancing, enlarging, and reenforcing the soldier's masculinity.[47] In a revelatory study, Paul Achter draws attention to "The Rhetorical Domestication of Twenty-First-Century Veterans of War" by casting a critical eye on the "'geographic' and 'embodied' phenomenon" of understanding the politics of war through the veteran's body (both injured or not). In Achter's words, "bodies are fluctuating signifiers whose forms are molded and shaped by a range of cultural forces, including medical technology, personal desires, and public discourses where definitional arguments about bodies find expression."[48] If "war is injuring" of human flesh, then war itself becomes a "fluctuating signifier" in which the bodies' "health, their death, and their wounds serve as a metonym for both the nation's health and for the condition of the war."[49] In other words, a narrative of war focused on the integrity of the nation-state involved in armed conflict necessarily glosses over the implications that the wounded, maimed, and injured soldier has for an understanding of war as an efficient, rational, and objective affair. The injured veteran's body does not fit within the mainstream corporeal narrative of strength, valor, and invincibility but becomes mired in a discourse pitching "home" bodies versus "enemy" bodies, "strong" bodies versus "deficient" bodies, "healthy" bodies versus "injured" bodies, "victorious" bodies versus "dead" bodies, the latter in each pair informing vulnerability and constructed weakness as an underlying feature of the human condition. Children's dead "war bodies" make this realization even more urgent. What is more, the injured veteran/soldier body opposed to the injured/dead civilian body does something unexpected: civilian bodies become "war bodies," defying the weakness ascribed them. In the case of children, and because of their doubly vulnerable status, as well as civilians and minors, the drive to protect evokes the need to valorize the child's body.

The injured soldier's body, nevertheless, still exists in the realm of the public imaginary as heroic, despite its less than masculine appeal and despite the shocking scars of combat. It is produced, for good or bad reasons, by the war machine as the nation-state and, as such, exists as a witness, a living testimony to the destructive power, reach, and omnipotence of armies. Amputee veteran runners, for example, epitomize the resolve to prove "that injuries do not prevent them from doing the things they could before,"[50] thus making sense of war not only through the lens of the soldier's resolve but also helping to make sense of war for civilians as a universally applicable and personally meaningful experience of defying weakness. The officially produced nature of the cause of the injury is thus obscured, turning an uncritical eye to the "war machine" and its collateral damage, as we zoom in on stories of personal recovery and perseverance instead. Soldiers' bodies are valorized both during and after combat as perfect representations of a morally superior, strong, commanding narrative of the nation-state. Yes, "injury must at some point be understood individually because pain, like all forms of sentience, is experienced within, 'happens' within, the body of the individual."[51] How do children's dead bodies make this an even bigger imperative?

CHILDREN, WAR, AND INTERNATIONAL LAW

Like experiences of wounded veterans and dead heroes, the bodies of dead or murdered children figure as a shocking spectable that escapes lasting attention for the following two reasons. On the one hand, because children are not supposed to die a violent death at war, the incidences of their deaths are explained as an exceptional occurrence. Second, because they are not the usual combatants we are used to evoking when discussing war, their ambiguous position on our screens and in our minds almost deprives them of the focused and committed attention they deserve to untangle their "no-man's-land" kind of status. Indeed, what is one to make of children's dead bodies during conflict, specifically with attention to the death of children fleeing oppressive, violent, and discriminatory regimes? In the cases of Aylan Kurdi (Syrian) and Fareed Shawky (Yemeni), their bodies cannot be read simply or exclusively as trophies of war, nor can they be understood as straightforward sacrifices for the nation, because they were not involved in an organized and institutionally supported struggle of any sort. Beyond the question of how to understand their deaths in general and their lives more specifically, the question that interests me here deals with the relationship of their deaths to IR and its program of arresting insecurity and reinventing itself in the process. If the body of the wounded veteran alludes to a kind of valor and pride originating from an association with the nation-state, and if the body of the dead soldier evokes feelings of pride and nationalism, do the dead bodies of children at war uncover a vulnerability in the nation-state, even a crisis in the ability of the latter and its narratives to justify and explain death away from loss, disorder, and tragedy? What about the failure of the responsibility to protect those perceived to be the most vulnerable?

Here are the questions I pose in this chapter: If war is the embodiment of the spirit of the nation-state writ large across the bodies of its soldiers, and if war in its dailiness has almost ceased to shock, then in what important ways does the body of Aylan Kurdi, for example, challenge the self-understood and self-evident normality of war in IR? Additionally, if injured veterans can be reconciled with anticipated collateral damage resulting from "just wars," their postwar physical setbacks explained as the usual consequences of war(s), how do the dead bodies of children unsettle a narrative of wholeness and integrity of the nation-state as such? Moreover, if a dead child is a consequence of such a war, does that raise questions about whether the war was just? I argue that the modern nation-state is a repository of images, ideas, and ideologies that aren't determined or fixed but are constantly recreated and struggled over and as such, images of dead and dying children's bodies do belong to the IR narrative while, at the same time, bearing witness to the IR's own silences and omissions.

Following from Judith Butler's attestation that only certain bodies matter during war, only certain bodies count and are counted, whereas others are not recognized as fully human, becoming unintelligible and "ungrievable,"[52] I argue that the dead bodies of these nonwhite children symbolize an individual's "ungrievable" relationship to the larger register of national/official discursive constructions of war.[53] In this way, they resemble Agamben's *homo sacer* even though their deaths do not quite enter the register of exceptional sovereignty run amok.[54] Paralleling my discussion in chapter 2 of trafficking survivors exceeding "the state of exception," so here the exception becomes the norm as killing in times of war is challenged by the lack of logical, reasonable, and acceptable explanatory possibilities that kids washed on shores offer to the sovereign center's position as security providing. What is confirmed once and for all is that "the moment of death nevertheless serves as a graphic example of forms of oppression."[55] Ironically, this is evident also when looking at the Convention itself.

The general rights enumerated in the Convention are rights to provision, rights to protection, and rights to participation.[56] Some of the specific rights include a right to parental guidance (Article 5), a right to life (Article 6), a right to a name (Article 7), a right to a nationality (Article 8), a right not to be separated from his or her parents against his or her will (Article 9), a right to be reunited with parents and, whenever possible, maintain contact with parents (Article 10), a right not to be transported and kept abroad for illegal and criminal purposes (Article 11), a right to participation (Article 12), a right to freedom of expression, thought, conscience, and religion (Articles 13 and

14), a right to freedom of association (Article 15), and various other provisions ensuring that in the absence of parents or in the absence of parents' ability to care for their children, state parties can implement and put into effect all necessary measures to restore the child's security and safety. Andressa Gadda does well to point out the inherent contradictions between some of these articles providing, on the one hand, the right to protection and the right to participation without, at the same time, guaranteeing the right to know one's rights.[57] Important as it is, the Convention privileges state parties and their role as ensuring and guaranteeing rights for children, as well as privileging a particular view of childhood largely informed by Western conceptions of the meaning of life.

Next, I briefly overview some of the prevalent critiques of the Convention before I return to specific examples of the dead bodies of children. First, the Convention was a direct result of long-term cooperation among nation-states and, as such, was/is predicated on the existence and prevalence of state structures. As a result, "the creation of the modern state and national culture is integrally related to the creation of new sorts of gendered, and age-graded subjects and spaces, and the establishment of institutions variously engaged in spreading these constructions throughout society."[58] Gadda traces the history of the construction and understanding of "childhood" as a coherent fixture in discourses of power and social control, as well as its investment in a liberal ideology of rationality, stability, and hierarchical relations of power.[59] In order for children to be ready to assume constructive roles in society, their preparation for adulthood and their rearing have become about more than just survival. "Thus, the child, like the primitive inhabitants of the colonies, had to be introduced to the liberal doctrine in order to become fully rational and, consequently, eventually gain the rights to full citizenship," meaning that "children had to be under greater surveillance of responsible adults."[60]

This led to the normalization and standardization of educational requirements, which themselves did not necessarily impose sweeping corporeal punishments on children but did something even more effective: they disciplined them into adopting behaviors deemed acceptable, normal, productive, and rational. In the words of Bertrand Ogilvie, the assimilation of an individual into universal morays and norms is never simply the acquisition of the values of humanity that contain power but is always a domination that needs to be "enforced" through the body, leading to a state of voluntary servitude.[61] Said processes of normalization, in turn, aim at reducing the complexity and ambiguity inherent in growing up and thus prepare children for becoming active agents of reducing said ambiguity and complexity as adults themselves. To put this another way, "The acts of power and knowledge that govern our expectations about students' abilities and behaviors are characterizations unique to the modern Western style of knowledge—even though this way of thinking is so entrenched that it is almost impossible for most of us to imagine how else a child could be approached."[62]

Children's rights merit engagement if only so as to shed light on the evolution of the concept and the movement dedicated to its proliferation. Despite several prominent examples of children protesting either poor working conditions, Apartheid, or educational reform before the 1970s, and even though the nineteenth and early twentieth centuries offer plenty of examples of campaigns and initiatives aimed at "improving" the lives of orphans and child workers, it was not until the late 1970s and early 1980s, following the adoption of the International Covenant on Economic, Social, and Cultural Rights and the International Covenant on Civil and Political Rights in 1966, the Convention on the Elimination of All Forms of Discrimination against Women in 1979, and the Convention on the Rights of Persons with Disabilities in 2006 that discussions regarding rights of children came to be.[63] It is important to note that the first Declaration on the Rights of the Child was adopted in 1934 by the League of Nations and was, like many similar documents, the direct result of "the unprecedented suffering of children during World War I."[64] As with similar documents, negotiations over the specific text and provisions of the Convention on the Rights of the Child (1989) lasted for more than ten years, including the contributions of member states as well as various nongovernmental organizations.

Two of the most contentious issues faced by the parties involved in negotiating the current Convention had to do with the definition of a child and with children's service in armed services. It should be noted that though the United States of America still has not ratified the Convention,[65] it was active in proposing seven articles to its text, attributed by some to its desire to outdo the Soviet Union in the context of the Cold War.[66] In 2000, the United Nations adopted two optional protocols relating to children's involvement in armed conflict, the sale of children, child prostitution, and child pornography.[67] Additionally, the Rome Statute setting up the International Criminal Court, which the United States also has not recognized, offers important provisions relating to children's rights with respect to their treatment in times of war, covering things like "conscription, enlistment, or use in active hostilities of children under the age of 15; intentionally attaching schools; and rape and other sexual violence."[68]

In the context of the Convention rights, education and citizenship are enmeshed and superimposed on one another as they work to construct identity through processes of education as governing aimed at maintaining the status quo and the social order.[69] As a result, what used to be conceptions of childhood rooted in an idea of children as savages evolved into ideas of children as vulnerable, indefensible, sacred, and malleable. The Convention, in this sense, is clearly the product of such engineering and can be read in the context of practices and mechanisms of control and governing intended to produce a docile, obedient, and productive liberal subject, defined as such not so much because of his/her right to have rights but because of his/her right to protection by the nation-state.[70]

In summary, some of the more common critiques waged at the Convention have brought to light its focus on the rights of adults to shape children in their image, its fundamentally permissive humanitarian provisions that dictate that in cases of violations of children's rights, interventions to save said children are justified and desirable. Additionally, a rights-based discourse informing the Convention argues for rights in terms of needs where "the child has a right to what she needs in order to have a childhood as envisaged by Western societies.[71] Additionally, critiques waged by African states more specifically pointed to the highly Western nature of the language and provisions of the Convention, in which African countries were underrepresented and whose children have fundamentally different political, economic, social, and cultural realities.[72] A critique related to this chapter is that the Convention does not provide many protections for children from violent conflict, which is only partially addressed by the Rome Statute and the Just War Doctrine, nor does it establish for them a special status, as minors, in the registers of warring nation-states. Thus, while the discipline and upbringing of children is supposedly closely monitored by various state apparatuses, like teachers, psychologists, legal guardians, social workers, and parents, the state has done little to clarify what its position is when it comes to its obligations to protect the lives of children at times of war. This is at the heart of children's vulnerability in times of war, which is underscored by human vulnerability at war, raising the bar for engaging with "war bodies" in general.

AYLAN KURDI

> One of the essential characteristics of modern biopolitics (which will continue to increase in our century) is its constant need to redefine the threshold in life that distinguishes and separates what is inside from what is outside.
> —Giorgio Agamben, *Homo Sacer*, 131

The death of Aylan Kurdi was publicized and grieved around the world, and various testaments to his memory have since been erected, either in the staged performance on a Moroccan beach in which adults dressed as Aylan Kurdi assumed the same position he was found in, or the giant sand statue of him on a Pakistani beach representing his final moments in that almost life-like, sleep-like position in which he was found.[73] Aylan's body, when discovered on September 2, 2015, seemed at peace—his red T-shirt and blue shorts reminiscent of the bright colors of children's clothes, and his position reminiscent of that of a sleeping child. There was no blood on his body, as if he had been washed preburial as is the tradition in most cultures, no traces of violence or conflict, and his body was delivered onto the Turkish shore whole, beautiful. His dead body does not bear the traces of other dead bodies at war at first, except for its lifelessness. The feeling of "childhood" his body communicatesd was intensified when looking at the Turkish aid worker, a tall man indeed, who was photographed carrying him off the beach. Like a parent would carry a child to bed or when hurt, the man holds Aylan's body, little arms and feet dangling. Aylan's body dwarves even more in comparison to that of the tall Turkish man.

Alongside Aylan, his older brother, Ghalid, and his mother also died, though their bodies were not washed ashore and thus pictures of their bodies weren't available. In a way, his death overshadowed theirs; the tragedy that he represented simply couldn't be matched by two more drowned, but not displayed as such, bodies. The sensation that was Aylan's body, seized on by the media, displaced the attention that should have been given his mother and brother, too. It is important to note this representational slip. Could it be said they did not suffer a violent "war death" because their bodies were not treated the same way? Or could it be that their bodies were not treated the same way before they did not suffer a more "violent," public death? Following from my discussion in the opening section of this chapter regarding the face of the wounded soldier and the challenge his/her disfigurement poses to our engagement with him/her, Aylan's mother and brother were absent because they did not capture the public's imagination in a way sufficiently shocking as to propel them onto our computer screens and newspaper covers, much like the 200,000 Iraqi children Judith Butler talks about. Their deaths were omitted alongside those of other drowned migrants whose dead bodies, because outside the public gaze, also recede into oblivion. Who will pay tribute to the normalized death on the way to "Fortress Europe," where the free movement of people is heavily regulated, limited, and militarized?[74]

So what is it about Aylan's body that marks his death as different compared to the thousands of other migrants dying in an attempt to cross the Mediterranean or Aegean Seas in all kinds of boat-like devices? How is his "war body" inscribed with political importance even though, in the end, his death did not lead to significant changes in EU and U.S. migration policies toward incoming migrants? "Ungrievable,"[75] Aylan as a child, a migrant, a "body" of war, and a human being exists outside the clear-cut registers of policy discussions in which he is treated as an accidental, if unfortunate and very tragic, victim of war. In the aftermath of his death, traffickers were detained, their practices of overloading inflatable dinghies were condemned, the unreliable life vests that were given migrants were scrutinized, and other details of technical nature, not forgetting the weather, became the public culprits in this instance.[76] In the more than three years since his death, stories of overflowing dinghies and migrants drowning without life vests have come in through the wire regularly, the half-life of shock wearing off to welcome another sensation, a government shut-down, elections, and children dying in detention on U.S. soil. Turning to the Convention again, Aylan's right to protection and to life could not be ensured precisely while in the care of his parents, but not because his parents were not caring for him. Despite the missing information about the specifics of his drowning, the sneakers in the picture of him on the beach must have been lovingly put on by a parent that same day in the belief that he was going to safety.[77] Just like I put my children's shoes on their feet before they leave for school, just like I had surely done on that September 2 in 2015, so must his parents have helped him get dressed for the day. His parents were, themselves, in a vulnerable and precarious situation in the waters of the Mediterranean, where his and their bodies existed rightless. One could say these migrants were "just bodies" whose right to death took precedent over their right to life.

In the context of traversing an international maritime border, it appears as if nation-state control over the provision of rights to individuals is suspended. Neither was the state there in the open water to protect people in boats, whether as legal or illegal migrants, nor was it there to apprehend them prior to arriving in national waters. In a sense, the migrants' right to travel was guaranteed only insofar as they were free to leave a country but not regarding their freedom to arrive in one. The Sea thus provides an example of state powers that exist because they are imagined and not because they are actually present. Because the Sea stands between Greece and Turkey, Turkey and Syria, Libya and Italy, it becomes a state of its own, a territorial demarcation of belonging and nonbelonging at the same time, an in between space, a zone of indeterminacy, where sovereign and international law are blurred, where moving water and currents mock the fixed territorial preoccupation with lines on a map, and where the human body enters and is defined in flux. The upward and downward rocking of boats, the seemingly endless expanse of water, the lack of territorial markers on the horizon, and the parts of the human body that are, literally, uprooted, floating, restless while traversing the Sea bear witness to an insecurity that is foundational to the workings of the state and its justifications for removing "others."[78]

Comes Michael Dillon with the question, "But who is not a stranger? What human is not a mortal and temporal being bearing the mark of difference within itself as that which makes itself possible, difference which does not merely individuate each human being as the individual self that it is, but which each actively bears constitutively within itself?"[79] Comes Bertrand Ogilvie bearing witness "*aux limites*," to the limits of representing instances of extreme violence.[80] Aylan's death, in this sense, enters the register of a violent "war death," which discloses the fact that even though our bodies are ours to care for and use, "the very bodies for which we struggle are not quite ever only our own."[81] Aylan's lifeless body on display creates an imperative to make "grief as part of the framework within which we think our international ties."[82]

It is significant that Aylan existed, and still exists, inside but also outside the registers of IR. He does not belong to any one register that is the product of and responsibility of national governments. He inhabits a no man's land and as such lives beyond the date and facts describing his end. He does not belong neatly either to the Syrian resistance, the Syrian government, the humanitarian effort, or to IR but belongs to them all at the same time, yet also exceeding them. His death is an issue of social, ethical, humanitarian, political, and legal concern, yet he himself, a child of three, was not a member and a legitimate participant in either of these registers. His death is a tragedy and an accident but also an example of negligence, the result of state-sponsored violence, and the consequence of the machinations of criminal networks exploiting his family's vulnerability. They could have crossed the Sea had they been in a better boat, without the threat of drowning, and thus the threat that loomed over them was man-made and avoidable. Unfortunately, good customer service is reserved for wealthier migrants or, even better, for paying tourists and Aylan, who was a "vagabond,"[83] disposable and "ungrievable," had no access to them. At three years of age, his powerlessness and rightlessness represent the ultimate referent object of power as well as a powerful critique of state narratives of order, childhood innocence, and the romanticization of youth. He did not come from a largely depoliticized world of play and carelessness and, as a result, his death is a caricature of youth as innocence. Because of that, it will shock forever. He is the ultimate "child of war" because the only way we know him remains firmly grounded in his death.

In 2015, 3,771 people came to their deaths crossing the Sea, and Aylan, his brother, and mother were three of them.[84] In 2016, that number rose to 5,079[85] despite the number of people who made the dangerous crossing dropping from 1,015,078 in 2015 to 363,348 in 2016.[86] For 2017, the numbers show the number of migrant deaths at 3,119 out of 178,721 arrivals.[87] This puts the proportion of deaths versus arrivals for 2017 at 1.5 percent and in 2018 so far, at the time of writing this in early February, that proportion is already at 5.2 percent in which out of 6,624 arrivals, there have been 343 deaths.[88] In 2018, the number of migrant deaths in the Mediterranean came up to 242, and in 2019, in the month of January alone, this number stood at 217.[89]

Aylan's death, what with his colorful clothes, chubby legs, and semi-fetal position, faces the spectator with an utter dissonance of appearance and reality, underwriting a problematic tension in any attempt to explain and/or engage with his death in a meaningful, rational, appropriate, and nontrivializing way. This difficulty is precisely the lesson: death is difficult, final, and not justifiable, though Aylan does look like he was on his way to the playground. The conclusion that remains, though not a satisfactory one, returns me to the work of Judith Butler and her brilliant and insightful discussion on mourning, trauma, and violence. To realize that "certain lives will be highly protected, and the abrogation of their claims to sanctity will be sufficient to mobilize the forces of war" while "other lives will not find such a fast and furious support"[90] is also to insist that our "'common' corporeal vulnerability"[91] presupposes an ethics of encounter. It is in that space that we meet Aylan for the first and last time, his "primary vulnerability"[92] presupposing that if "I am nowhere without you"[93] as my other, you are also nowhere without me.

Departments of Defense and State personnel, standing in for the nation-state, know how to deal with the bodies of dead veterans covered with the American flag. "The co-opting of the dead into national narratives" thus takes a literal form, as "fallen soldiers remain the property of the state. Having been sanctified or made sacred, they are no longer 'ordinary' people."[94] Edkins further refers to the rows and rows of military graves in Arlington who not only inscribe the fallen as part of the narrative of national pride but also inscribe their families, seeing how their wives' names (but not those of their parents or other relations) are written on the reverse of the gravestone.[95] The ambiguity that accompanies military deaths, whether they are sacrificial or not, demands an engagement, but an engagement that doesn't explain violence as a necessary part of combat, especially when deaths at war are those of children.

IR can offer a sensible, rational explanation for the prominence of terrorist and intra-state conflicts around the world, having originated narratives of globalization, ethnic conflict, and racial differences as the new "clashes of civilizations." IR has even opened its doors to some gendered accounts of power, recognizing the importance of engaging alternative voices, if only to silence them in the process.[96] But can IR address the dead "war body" of Aylan Kurdi? It cannot. Yet his death is anchored in IR's logic, which buttresses states' concern with national security and where the immigrant is defined and understood in the shadow of the terrorist, the criminal, and the outsider. Their belonging is jeopardized simply by virtue of their "cultural and racialized markers 'as visual evidence of cultural inadequacy'"[97] in a Panopticon Europe[98] preoccupied with what Rosière and Jones have termed *teichopolitics*, or the "politics of building barriers on borders for various security purposes."[99]

In other words, while Aylan Kurdi's death became widely publicized and his small lifeless body a symbol of the refugee plight, his death is unsettling. Subsequent deaths of children[100] trying to cross into Turkey, Italy, Greece, and Spain in boats received much less attention, as if the originary moment of shock quickly turned into another "tragic but expected" accident, as more and more refugees pour into inhospitable Europe. These deaths have receded into the background as if the half-life of the terror, outrage, and trauma that followed Aylan Kurdi's death was the mere length of a news cycle, maybe two. How could any other death outdo Aylan's?

Unable to arrest the flow of Syrian, Afghani, Yemeni, Libyan, and other refugees, the world watches from the sidelines as they are beaten, discriminated against, persecuted, imprisoned, cheated, deported, and left to die. The modern nation-state and its logic of order and territorial and legal authority thus populate a graveyard filled with the bodies of the "unwanted," with Bauman's "wasted lives." Not necessarily conjugated as the excess of the modern capitalist project, as in Bauman's example, these refugees nonetheless represent excessive humanity, humanity doomed, Agamben's *homo sacer*, though I here argue that Aylan's death is neither simply a murder nor a sacrifice[101] but both, as the likes of him exist in a register of constant vulnerability, serving an ongoing sentence predicated on their origin and country of birth and generated and sustained by demarcations of inside and outside, self and other that bring forth their inevitable destruction.

Aylan's death, and the deaths of the thousands of other migrants crossing borders, stands in for other instances of exterminations and extreme violence in the world. The Turks killing Armenians, Stalin's gulags, the Nazis gassing millions of Jews, Gypsies, the infirm, and political prisoners, the Tutsis chopping up the Hutus, the Hutus murdering Hutus, Pol Pot murdering its own in the thousands, the Janjaweed genocide in Darfur, the ongoing persecution of the Rohingya in Myanmar—the list goes on—all coalesce and are represented in the body of Aylan. But about this there is silence. What we hear is the nation-state, having already created and justified the need for "the state of exception" as an organizing principle of modern government, haggling over quotas, press statements, and memos regarding who bears responsibility for the tragedy unfolding before our eyes. In the end, an emphasis on a "politics of containment" gives governments the right to claim, as in the case with human trafficking discussed in chapter 2, the right to use "the most repressive measures that appear to violate the very spirit of democratic traditions" in which migrants exist in extra-legal extra-territorial zones, "flaunting respect for human rights."[102] The violence underwriting the revival of militarized nationalist border policies can be seen both on the inside and the outside of states, the example of Fareed Shawky's death standing in for the former.

FAREED SHAWKY

Alongside Aylan comes Fareed Shawky, a six-year-old boy who died in Yemen in 2015. Before I proceed, I share an excerpt from a news story relating his death.

> Lying tensed on his hospital bed, a six-year old pleaded with his doctors. "Don't bury me," he implored, "don't bury me." Caught on camera, Fareed Shawky's terrified whimper was shared online more than 50,000 times. In the video, doctors laughed gently with one rubbing the child's knee, as if to reassure him he would be OK. But a few days later, Fareed's family would bury him, counting the boy as one of the more than 500 child casualties of Yemen's war. He suffered deadly injuries in a missile attack launched by Houthi rebels on October 13 as they fought for control of his besieged hometown of Taiz. His dying plea gave voice to the fear felt by millions of Yemeni children caught up in their country's war.[103]

What is shocking and arresting about Fareed is not so much his death but his fear of dying before that death. In the video showing the incident described unfold,[104] he is begging his father and the doctors not to bury him, not to put him in the ground, as if somehow, in life, he had already experienced the horrors of death, as if, somehow, death had already put its grip on him.

Fareed's death was not unique or extraordinary, but there was something about speaking to/about his death before it happened, anticipating the disappearance of the body in the ground, that brought to the fore the reality of all children dying in wars and not only the ones caught by cameras. His plea not to be buried (forgotten, removed, ousted, discarded), his absolute horror at his impending death, at what happens to the body in death, at what would happen to his body has become the mark of so many other children caught in a conflict they cannot escape. Yet Fareed was able to voice his fears, to send a last plea into the world, when Aylan was unable to do that. To bear witness to one's own death before it happens, at six, is nothing short of exceptional and this exceptionality calls forth a witness, a political recognition.

It is difficult to watch the video of Fareed being treated for his wounds without feeling haunted by his words (which an English speaker does not understand without interpretation), his gestures, his eyes, and his body, which all speak for themselves. He does not look like he is about to die, yet he squirms, makes scared faces, and asks for assurance that he will not be buried in the ground. Neither his father nor the medics nor anyone else could protect him and assure him that he would not die, yet they do because he appears to be out of danger and the trauma that is visible is something they can treat. The dissonance this creates for the spectator between the reality of his dead body now and the video of his wounded body inaugurates a crisis of identification, even a crisis of empathy, both for the state and for Fareed. If a well-built and strong body is symptomatic of a strong state, then Fareed's shrapnel-covered body reveals the well-being and prosperous state as a myth. Fareed and his body are symptomatic of a vulnerable state and his inability to make sense of the violence around him is mirrored by his physical disorientation as his body is also unable to make sense of the world. Commenting on the body at war and the ugly realities of injured bodies, Vivienne Jabri attests that bodies "are not deemed to constitute the subject-matter of a mature discipline that concerns itself with the abstractions of the international system."[105]

Returning to the questions guiding the discussion in this book, more specifically my engagement with risk and (in)security as the underwriting principles of modern state power, Fareed and Aylan as the new bodies at/of war challenge us to extend our engagement with war and the international into an engagement with trauma as part of the fundamental stuff that makes up our discipline. The trauma I am talking about relates to the experiences of individuals faced with terror, violence, and injustice who have little recourse to defending themselves, individuals, like Aylan's mother and brother, who remain invisible both in life and in their deaths. It is children like Fareed and Aylan, together with countless others, that embody the passionless, frigid, narcissistic states that claim ownership of our bodies, our lives, and our feelings. William Butler Yeats in *The Second Coming* gives a chilling "diagnosis of the twentieth century" and one still applicable today: "The blood-dimmed tide is loosed, and everywhere / the ceremony of innocence is drowned; / the best lack all conviction, while the worst / are full of passionate intensity."[106] Aylan's "innocence drowned" washes on the shores of Panopticon Europe contoured and barricaded by fear, intolerance, and militarization. This should make us all pause. Take, for example, the victims of the April 4, 2017, gas attack that occurred in Khan Sheikhoun in rebel-held Idlib, Syria, where a total of eleven children and sixty-one adults were reported dead, with up to two hundred others injured.[107] Or take the August 17, 2017, example of the shocked Syrian boy and his sister seen in an ambulance, after nighttime bombing in their neighborhood claimed the lives of five other children and about a dozen adults. It is to him that I turn next before concluding.

OMRAN DAQNEESH

His name was Omran Daqneesh and in 2017 he was five. The image that made him "famous" and, I argue, a political being/subject for the first time in his life sees him sitting in the back of an ambulance in Aleppo, covered from head to toe with dust, blood dripping from his head.[108] His face is expressionless, defying intense emotion in the midst of the violence going on around him. He looks stunned, dazed, and is sitting absolutely still without uttering a word, only casting a sidelong glance around him from time to time, as if in disbelief. He is not screaming, though he is injured, and he does not make any indication that he understands what is going on around him, at least not in words. His body—dirty, injured, and still—speaks volumes. Around him, the work of death[109] is going on. Omran is small, his hair is disheveled, the picture is taken in the middle of the night, and the artificial lights of the ambulance where he sits are jarring, too bright, almost offensive. They erase the darkness around him, but also the shadows, and what we do not see in the ambulance is precisely to what Omran's body bears witness. There are explosions outside, the dust covering his body is the stuff of buildings but also probably the stuff of other people's bodies. Were he scared or surprised, in the ways people are scared and surprised when something unexpected and violent happens to them, his body would be animated, he would be asking for his parent, crying, or trying to escape. That would almost be a better, more fitting reaction that would also save the spectator from experiencing the horror lived by Omran. But he does neither of these things. He is the epitome of the "new face of fear." He seems numb, indifferent almost, or just so stunned as to be unable to process what is going on around him. He is traumatized, yet he occupies a universe in which the "traumatic event . . . is meaningless,"[110] not carrying any productive meaning with it, any truth, and not something to which we can unproblematically attribute meaning. Even though what happens does not make sense and never will, which is why one cannot easily bear witness to it in narrative form, he remains "stored there in your eyes."[111]

In the immediate aftermath of the trauma experienced, Omran does not speak. He inhabits the register of traumatic events that "are so shocking, so outside our expectations, that we do not know what happened," and so they are "indeterminate."[112] Jenny Edkins makes the point in her *Trauma and the Memory of Politics* that memories of traumatic events are often "detemporalized" because they are not allowed to manifest themselves as ongoing, yet in the cases of these diseased and injured children, I believe we can add a further layer to her discussion. The camera has frozen Omran, known as "The Face of Aleppo,"[113] and even though he appears as if he does not know what is happening, I believe his frozen image stands in for the memories and traumatic experiences of Aylan and Fareed too. What Aylan and Fareed are unable to express because they are dead—their fear, their disappointment, their loneliness, their disbelief—Omran communicates. Dead children speaking for the living, and the living sitting in for the dead. Against linear, historical, narrativized time that has "beginnings and ends"[114] comes trauma time, which is experienced belatedly, in which traumatic recall takes control of the speaker, who "cannot control the way the trauma is recalled: it is often related slowly, in real time."[115] Trauma time, though shaped by violence, war, injury, and terror, also does something unexpected: it exceeds the "work of death" as the work of securitizing the subject and ushers in a realization that "loss has made a tenuous 'we' of us all"[116] in which grief, suffering, and pain are the ultimate political experiences of our times and in which "we're undone by each other. And if we're not, we're missing something."[117] Trauma thus always rewrites politics as usual.

To step away from Omran's image, from the picture of Aylan on the beach or the video of Fareed asking not to be buried, is betrayal, cowardice. So mourn them we must as our own children. Judith Butler offers the insight that in order for "a body" to be mourned, it must necessarily belong to someone who is considered "someone," "a person," "a life," which refugees, foreigners, terrorists, and the disposables are not. So "if 200,000 Iraqi children were killed during the Gulf War and its aftermath, do we have an image, a frame for any of those lives, singly or collectively? Is there a story we might find about those deaths in the media? Are there names attached to those children?"[118] And if there is no remembering, no obituary, does this mean these lives have never been? One thing is certain, they are not grievable in the same way in which the lives of fallen American servicemen are publicly grievable. In the instance of Omran, there was plenty of discourse in the public eye about his image, plenty of outrage. Yet his silence in the face of the trauma makes it hard to "know" him. He draws us in even though he does not cry out so that we may distinguish our cries from his; he does not stand up so that we may separate our grown bodies from his; he does not seem scared so that we may comfort him. He just sits there, drawing us in, haunting us, reminding us that "the body implies mortality, vulnerability, agency: the skin and the flesh expose us to the gaze of others, but also to touch, and to violence, and bodies put us at risk of becoming the agency and instrument of all these as well."[119] The kind of trauma that Aylan Kurdi, Fareed Shawky, Omran Daqneesh, their families, and the families of millions of others experience daily transforms the discussion about the nature of risk, security, and war from one about military capabilities and preparedness to one about everyday suffering and having to reconcile ourselves with the images of drowned toddlers, shrapnel-covered six year olds, and injured, speechless five year olds.

A Google search for Omran Daqneesh in 2018 shows him in the company of his parents, his father laying praise on the Syrian government. At first, I found it hard to reconcile this with the violence done to him on that night in 2017. Surely, his father couldn't have been supporting what seemed at the time the forces that had inflicted the violence on Omran and killed his ten-year-old brother as a result. In fact, his father is recorded as saying, by way of critiquing the Western media, "They wanted to trade in his blood and published his photos."[120] It is unclear whether this statement was made to protect himself and his family, or whether it was representative of his true feelings. He is incensed by the fact that the Syrian Civil Defense units, known as "The White Helmets," had taken his children to the hospital without his permission.[121] Omran is seen, in the pictures accompanying the story, sitting quietly next to his father, quiet again. He looks healthy and no blood is streaking from his face. The intricate politics of the Syrian conflict overshadow his story, again, and warring parties are using him for their end goals.

CONCLUSION

As I put the finishing touches to this chapter, pictures of children covered in blood, ashes, and screaming in pain stream on my computer stream. It is not Aleppo, it is eastern Ghouta this time. The death toll from the past week is estimated to stand at 500, maybe even 600.[122] But who is counting? And what does it all mean? One picture of a boy being carried by a "White Helmet" arrests my attention. He looks to be no older than seven. As he is carried, his mouth is open in a scream but I cannot hear it. I am not sure if he is in pain, injured, or upset because someone he loves has just died. The picture does not tell the whole story. But his scream stays with me for days. It looks raw, though probably silenced by the roar of war around him. I want to know so much but am left with his wide-open mouth, his face disfigured in a grimace that a parent knows only too well—one of pain, discomfort, fear. He is yet another face of war, another "war body"—they are silent, loud, dead, alive, all at the same time. They keep coming, streaming from the computer screen and invading our world.

Fareed's plea not to be buried reminds us that war is a fundamentally emotional experience in which the luxury of "*not feeling*" is out of the question. Aylan and Fareed, for example, offer an insight into the fluid, broken up, and contradictory nature of IR's engagements and pronouncements on war and people's actual experiences of conflict in which an engagement with people's everyday bodily experiences depict "war as a comprehensive whole that has a teeming life alongside and sometimes in defiance of what statesmen, militaries, strategists and IR specialists say about it."[123] In such a way, politics is revealed as an emotional experience as well and ought to be engaged as such first and foremost. Narrating the body is imperative as a methodological tool for engaging with atrocities, a tool for making visible experiences of war and for uncovering the workings of power and control as social regimes. Yet the narratives in mind here are not linear, nor are they chronological. They do not "produce stable and reliable signifiers" as the "non-representational comes to the fore."[124] Aylan's death and Fareed's plea "reveal the ghost of the materiality of the body, and thereby escape the variety of ways bodies are appropriated for the purposes of stabilizing the national identity."[125]

The lives of Aylan, Fareed, and Omran cannot enter our collective histories without reshaping our cultural memories, histories, and debates. In the same way that the act of creation (birth) instantiates certain productive relationships of power predicated on futuristic, hopeful, and possibilistic narratives, the act of termination (death) does not simply end these relationships but changes them in ways that face us with the nauseating nature of the experience that is war. A conversation about inside and outside, about inclusion and exclusion, about power and domination, about life and death, about suffering and emotions must necessarily figure in any account of IR, lest the discipline resign itself to irrelevance. The greater the indeterminacy in the conversations regarding these topics, the better positioned we are to address the experiences of children in all their fractured beauty and significance.

Fareed's body, in the hands of the medical personnel, becomes a symbol of war as such, a symbol not simply of its senselessness (for wars are fought and will be fought still) but of its sickness, brokenness, chaos—of the ways in which it has been and is terrorizing us all. Fareed's body bears witness to the singularly unrepresentable nature of suffering that is war, suffering that I argue is the most significant thing about war, suffering that IR does not engage with or know very well. We never see Fareed's dead body, as we do Aylan's, but we learn that he died a few days later and was buried by his family, his father standing in the distance, unable to watch his son's body being lowered into the ground Fareed was so scared of inhabiting. While the bodies of veterans and soldiers help civilians make sense of war-time violence, the bodies of children like Fareed and Aylan represent an unmaking in which war as experience and emotion exceeds war as knowledge.[126] In other words, only once we engage "war bodies" and their stories, pleas, and suffering can academics within IR begin to understand war, the core subject of our studies, as lived experience and politics as an emotional affair.

NOTES

1. Christine Sylvester, *War as Experience: Contributions from International Relations and Feminist Analysis* (New York: Routledge, 2013), 1.
2. UN Convention on the Rights of the Child, http://www.ohchr.org/EN/ProfessionalInterest/Pages/CRC.aspx (accessed January 4, 2018).
3. Kevin McSorley, "Toward an Embodied Sociology of War," *The Sociological Review* 62, no. 2 (2014): 108–9.
4. Tarja Väyrynen, "Keeping the Trauma of War Open in the Male Body: Resisting the Hegemonic Forms of Masculinity and National Identity in Visual Arts," *Journal of Gender Studies* 22, no. 2 (2013): 137–51; Tarja Väyrynen and Eeva Puumala, "Bodies of War, the Past Continuous, and (Ar)thythmic Experiences," *Alternatives* 40, no. 3–4 (2015): 237–50.
5. Sylvester, *War as Experience*; Jenny Edkins, *Trauma and the Memory of Politics* (Cambridge: Cambridge University Press, 2003); Thomas Biersteker and Cynthia Weber, *State Sovereignty as Social Construct* (Cambridge: Cambridge University Press, 1996); R. B. J. Walker, *Inside/Outside: International Relations as Political Theory* (Cambridge: Cambridge University Press, 1992); Michael J. Shapiro, *Politics and Time* (London: Polity, 2016).
6. Väyrynen, "Keeping the Trauma of War Open in the Male Body," 138.
7. Swati Parashar, "What Wars and 'War Bodies' Know about International Relations," *Cambridge Review of International Affairs* 26, no. 4 (2013): 618.
8. Achille Mbembé, "Necropolitics," *Public Culture* 15, no. 1 (2003): 12.
9. Ibid., 27.
10. Christine Sylvester, *Experiencing War* (London: Routledge, 2011), 129.
11. Zygmunt Bauman, *Wasted Lives* (London: Polity, 2004).
12. Donald Carter and Heather Merrill, "Bordering Humanism: Life and Death on the Margins of Europe," *Geopolitics* 12, no. 2 (2007): 248–64.
13. Ibid., 253.

14. Giorgio Agamben, *Homo Sacer: Sovereign Power and Bare Life* (Stanford, CA: Stanford University Press, 1998); Giorgio Agamben, *State of Exception* (Chicago: University of Chicago Press, 2005).
15. Judith Butler, *Precarious Life* (London: Verso, 2004).
16. Ibid., 34.
17. Trevor Noah, *Born a Crime: Stories from a South African Childhood* (New York: John Murray, 2016).
18. Parashar, "What Wars and 'War Bodies' Know about International Relations," 621.
19. I would like to thank Stefano Ballesteros for helping me work through this relationship.
20. Judith Butler and Athena Athanassiou, *Dispossession* (London: Polity, 2013), 83, 86, 99.
21. Hayley Halpin, "Children Suffered a Shocking Scale of Violence in Conflict Zones in 2017, UNICEF Warns," *The Journal.ie*, January 4, 2018, http://www.thejournal.ie/unicef-children-conflict-2017-3771913-Dec2017/.
22. Shashank Bangali and Sultan Faizy, "Record Number of Children Killed in Afghanistan Conflict in 2016, U.N. Says," *Los Angeles Times*, January 4, 2018, http://www.latimes.com/world/la-fg-afghanistan-report-children-killed-20170206-story.html.
23. Giorgio Agamben, *Remnants of Auschwitz: The Witness and the Archive* (New York: Zone Book, 2002).
24. Parashar, "What Wars and 'War Bodies' Know about International Relations," 623.
25. Elaine Scarry, *The Body in Pain: The Making and Unmaking of the World* (Oxford: Oxford University Press, 1985), 63.
26. Ibid., 64.
27. Debra Lennard, "Censored Flesh," *The British Art Journal* 12, no. 2 (2011): 25.
28. Ibid., 27.
29. Ana Carden-Coyne, "Masculinity and the Wounds of the First World War: A Century Reflection," *French Journal of British Studies* 10, no. 1 (2015): 1.
30. Ibid., 2.
31. Ibid., 3.
32. Jenny Edkins, *Face Politics* (London: Routledge, 2015), 165.
33. Suzannah Biernoff, "The Rhetortic of Disfigurement in First World War Britain," *Social History of Medicine* 24, no. 3 (2011): 666–85.
34. Suzannah Biernoff, *Portraits of Violence: War and the Aesthetics of Disfigurement* (Ann Arbor: University of Michigan Press, 2017), 4.
35. Matthias Reiss, "Bronzed Bodies behind Barbed Wire: Masculinity and the Treatment of German Prisoners of War in the United States during World War II," *The Journal of Military History* 69, no. 4 (2005): 480.
36. Ibid., 484.
37. Moustafa Bayoumi, "War Stories from Soldiers and Body Washers," *The Progressive* (2015): 45.
38. Dimitris Foteinos, "Images of the Body: The Greek Physical Education Curriculum Since the Second World War," *History of Education* 41, no. 6 (2012): 809.
39. Michel Foucault, *Power/Knowledge: Studies in Governmentality* (Chicago: University of Chicago, 1980), 100–102.
40. Ibid., 39.
41. Foteinos, "Images of the Body," 50.
42. Emily Russell, *Reading Embodied Citizenship: Disability, Narrative and the Body Politic* (Piscataway, NJ: Rutgers University Press, 2011), 109.
43. Ron Kovic, *Born on the Fourth of July* (Brooklyn: Akashic Books, 2005), 111.
44. Scarry, *The Body in Pain*, 61.
45. Ibid., 62.
46. McSorley, "Toward an Embodied Sociology of War," 108.
47. One need only look at Captain America as the symbol of the broken/imperfect soldier's body made anew: strong, invincible, noble, and fantastic. The pervasive ideological shift valorizing the sacrifice and selflessness of the warrior is underwritten by a parallel narrative, created universe that enters the everyday consciousness of its audience as real, authentic, and true.

48. Paul Achter, "Unruly Bodies: The Rhetorical Domestication of Twenty-First-Century Veterans of War," *Quarterly Journal of Speech* 96, no. 1 (2010): 48.
49. Ibid., 49.
50. Ibid., 61.
51. Scarry, *The Body in Pain*, 65.
52. Butler, *Precarious Life*; Judith Butler, *Frames of War: When Is Life Grievable?* (London: Verso, 2016).
53. Butler, *Precarious Life.*
54. Agamben, *Homo Sacer.*
55. Butler and Athanassiou, *Dispossession*, 111.
56. Priscilla Alderson, "UN Convention on the Rights of the Child: Some Common Criticisms and Suggested Responses," *Child Abuse Review* 9, no. 6 (2000): 439–43.
57. Andressa Gadda, "Rights, Foucault, and Power: A Critical Analysis of the United Nation Convention on the Rights of the Child," *Edinburgh Working Papers in Sociology*, no. 31 (2008): 1–13.
58. Sharon Stephens, "Children and the Politcs of Culture in 'Late Capitalism,'" in *Children and the Politics of Culture* (Princeton, NJ: Princeton University Press, 1995), 15.
59. Gadda, "Rights, Foucault and Power."
60. Ibid., 6.
61. Bertrand Ogilvie, *L'Homme Jetable* (Paris: Éditions Amsterdam, 2012), 12; author's translation.
62. Gail McNicol Jardine, *Foucault and Education* (New York: Peter Lang, 2005), 9.
63. Jo Becker, *Campaigning for Children: Strategies for Advancing Children's Rights* (Stanford, CA: Stanford University Press, 2017), 6.
64. Ibid.
65. It is important to note that one of the main reasons that no U.S. administration has brought the Convention to Congress for ratification has to do with U.S. fears that the Convention takes away rights from parents to discipline and raise their children saying, fundamentally, that the rights of adults (parents) to deal with their kids as they want to, especially informed by religious reasons, takes precedence over the rights of children to enjoy rights as agreed upon and ratified by all the UN member states.
66. Becker, *Campaigning for Children*, 8.
67. Ibid., 9–10.
68. Ibid., 11.
69. Gadda, "Rights, Foucault, and Power."
70. Hannah Arendt's "Perplexities of the Rights of Man" offers a succinct exposé of the ways in which human rights as such, seemingly deriving from one's status as a human being, hold an empty promise unless protected and provided for by nation-states, ending up equating universal human rights with the rights of citizens. Peter Baehr, *Hannah Arendt, Totalitarianism and the Social Sciences* (Stanford, CA: Stanford University Press, 2010); Hannah Arendt, *The Portable Hannah Arendt* (New York: Penguin Classics, 2003).
71. Gadda, "Rights, Foucault, and Power," 8.
72. Becker, *Campaigning for Children*, 10.
73. Julia Nèjè, "Artists around the World Respond to Tragic Death of 3-Year-Old Syrian Refugee," https://www.boredpanda.com/syrian-boy-drowned-mediterranean-tragedy-artists-respond-aylan-kurdi/ (accessed February 4, 2018). *The Independent*, "Refugee Crisis: 600 Children Have Died Crossing the Mediterranean in 2016," http://www.independent.co.uk/news/world/europe/refugee-crisis-600-children-dead-mediterranean-2016-save-the-children-un-a7342131.html (accessed Feruary 4, 2018).
74. Amnesty International, *The Human Cost of Fortress Europe* (London: Amnesty International, 2014).
75. Judith Butler, *Precarious Life.*
76. Wayne Madsen, "Who Is Responsible for 3-Year-Old Aylan Kurdi's Death?" *Intrepid Report*, September 10, 2015, http://www.intrepidreport.com/archives/16226 (accessed February 24, 2018); *Time*, "Aylan Kurdis's Story: Behind the Most Heartbreaking Photo of 2015,"

http://time.com/4162306/alan-kurdi-syria-drowned-boy-refugee-crisis/ (accessed February 24, 2018).

77. Listening to a transcript of the *Time* story mentioned earlier, I was reminded of this by the words of Peter Bouckaert of Human Rights Watch, who was the first to tweet the picture for the world to see.

78. Tugba Basaran, "Security, Law, Borders: Spaces of Exclusion," *International Political Sociology* 2 (2008): 339–54; Wendy Brown, *Walled States: Waning Sovereignty* (London: Zone Books, 2010); Giorgio Hadi Curti, "From a Wall of Bodies to a Body of Walls: Politics of Affect/Politics of Memory/Politics of War," *Emotion, Space and Society* 1 (2008): 106–18; Carter and Merrill, "Bordering Humanism"; Stéphane Rosière and Reece Jones, "Teichopolitics: Re-Considering Globalization Through the Role of Walls and Fences," *Geopolitics* 17, no. 1 (2012): 217–34; Didier Bigo, "Security and Immigration: Toward a Critique of the Governmentality of Unease," *Alternatives* 27 (2002): 63–92; Dider Bigo, "Frontier Controls in the European Union: Who Is in Control?" in *Controlling Frontiers: Free Movement Into and Within Europe*, ed. Didier Bigo and Elspeth Guild (Aldershot: Ashgate, 2005).

79. Michael Dillion, "The Sovereign and the Stranger," in *Sovereignty and Subjectivity*, ed. Jenny Edkins, Nalini Persram, and Véronique Pin-Fat (Boulder, CO: Lynn Reiner, 1999), 118.

80. Ogilvie, *L'Homme Jetable*, 10.

81. Butler, *Precarious Life*, 26.

82. Ibid., 30.

83. Zygmunt Bauman, *Globalization: The Human Consequences* (New York: Columbia University Press, 2000).

84. "IOM Counts 3,771 Migrant Fatalities in Mediterranean in 2015," https://www.iom.int/news/iom-counts-3771-migrant-fatalities-mediterranean-2015 (accessed January 5, 2018).

85. "Mediterranean Migrant Arrivals Top 363,348 in 2016; Deaths at Sea: 5,079," https://www.iom.int/news/mediterranean-migrant-arrivals-top-363348-2016-deaths-sea-5079 (accessed January 5, 2018).

86. "Mediterranean Death Toll Soars, 2016 Is Deadliest Year Yet," http://www.unhcr.org/en-us/news/latest/2016/10/580f3e684/mediterranean-death-toll-soars-2016-deadliest-year.html (accessed January 5, 2018).

87. "Missing Migrants," https://missingmigrants.iom.int/region/mediterranean (accessed February 4, 2018).

88. Ibid.

89. https://missingmigrants.iom.int/region/mediterranean (accessed on January 27, 2019).

90. Butler, *Precarious Life*, 32.

91. Ibid., 42.

92. Ibid., 45.

93. Ibid., 49.

94. Edkins, *Trauma and the Memory of Politics*, 95.

95. Ibid., 96.

96. Cynthia Enloe, *Maneuvers: The International Politics of Militarizing Women's Lives* (Oakland, CA: Universisty of California Press, 2000); Cynthia Enloe, *Bananas, Beaches and Bases: Making Feminist Sense of International Politics* (Oakland, CA: University of California Press, 2014); Christine Sylvester, *Feminist Theory and International Relations in a Postmodern Era* (Cambridge: Cambridge University Press, 1994); J. Ann Tickner, *Gendering World Politics* (New York: Columbia University Press, 2001).

97. Sylvester, *War as Experience*, 219.

98. Godfried Engbersen, "The Unanticipated Consequences of Panopiticon Europe: Residence Strategies of Illegal Migrants," in *Controlling a New Migration World*, ed. Virginie Fuiraudon and Christian Joppke (London: Routledge, 2001).

99. Rosière and Jones, "Teichopolitics," 219.

100. *The Independent*, "Refugee Crisis."

101. Agamben, *Homo Sacer*.

102. Carter and Merrill, "Bordering Humanism," 256.

103. "Fareed Shawky, the 6-Year Old Yemeni Boy Who Becomes a Symbol of the War, Dies," http://www.telegraph.co.uk/news/worldnews/middleeast/yemen/11944336/Fareed-Shaw

ky-the-6-year-old-Yemeni-boy-who-a-became-symbol-of-the-war-dies.html (accessed February 10, 2018).

104. Video is available here: https://www.youtube.com/watch?v=DBimzkoJ5VM.

105. Vivienne Jabri, "War, Security and the Liberal State," *Security Dialogue* 37, no. 1 (2006): 47–64.

106. Slavoj Žižek, *Demanding the Impossible* (London: Polity, 2013), 85.

107. "Syria Chemical 'Attack': What We Know," http://www.bbc.com/news/world-middle-east-39500947 (accessed January 14, 2018); also https://www.nytimes.com/2017/04/04/world/middleeast/syria-gas-attack.html.

108. "Boy in the Ambulance," https://www.theguardian.com/world/2016/aug/18/boy-in-the-ambulance-image-emerges-syrian-child-aleppo-rubble (accessed January 5, 2018).

109. Mbembé, "Necropolitics."

110. Edkins, *Trauma and the Memory of Politics*, 36.

111. Michael Herr, "Dispatches," in *Unclaimed Experience: Trauma, Narrative and History*, ed. Cathy Caruth (Baltimore, MD: John Hopkins University Press, 1996), 181.

112. Edkins, *Trauma and the Memory of Politics*, 36.

113. Euan McKirdy and Mohammed Tawfeeq, "Omran Daqneesh, Young Face of Aleppo Suffering, Seen on Syrian TV," https://www.cnn.com/2017/06/07/middleeast/omran-daqneesh-syrian-tv-interview/index.html (accessed February 24, 2018).

114. Edkins, *Trauma and the Memory of Politics*, 40.

115. Ibid., 41.

116. Butler, *Precarious Life*, 30.

117. Ibid., 23.

118. Ibid., 34.

119. Ibid., 26.

120. Raf Sanchez, "New Photos Emerge of Omran Daqneesh," *The Telegraph*, June 5, 2017, http://www.telegraph.co.uk/news/2017/06/05/new-photos-emerge-omran-daqneesh-boy-became-symbol-aleppos-suffering/ (accessed February 24, 2018).

121. McKirdy and Tawfeeq, "Omran Daqneesh."

122. "Syrian Death Toll over 500 as Bombing in Eastern Ghouta Continues," *The Guardian*, https://www.theguardian.com/world/2018/feb/24/syria-death-toll-nears-500-as-eastern-ghouta-bombing-continues (accessed February 24, 2018).

123. Christine Sylvester, "War Experiences/War Practices/War Theory," *Millennium* 40, no. 3 (2012): 503.

124. Väyrynen, "Keeping the Trauma of War Open in the Male Body," 146.

125. Ibid., 147.

126. Christine Sylvester, "Experiencing War: A Challenge for International Relations," *Cambridge Review of International Affairs* 26, no. 4 (2013): 673.

Conclusion

Conclusions always bring about with them a sense of loss, of unconcluding. I would have liked to structure this conclusion around all the things this book did right and the reasons why it offered a novel and timely engagement with international relations' take on security, risk, and borders. But that won't be forthcoming. Instead I am reminded of the inception of this work, some years ago, and how far off course it has moved since then. The project was envisioned as a discussion of the ways in which borders work on the bodies of people who have experienced trauma during a border crossing. By borders, I envisioned those physical demarcations of territory I had so often crossed in my life, always legally. They have always had an ominous quality to them, and as such, they also fascinated me and seemed like a good topic for my first book. But even that did not hold. Over the years, in my imagination borders had almost acquired a mythical quality; they were "the things" that kept people as refugees, homeless, and displaced. Though I well knew better than to anthropomorphize what, in many instances, are invisible lines, they had fascinated me for so long that I had lost the ability to remind myself to see them for what they are—imaginary lines some times, border crossings other times, invisible markers yet other times. What this engagement with borders, security, and risk did bring to the fore were the various ways that nation-states experience insecurity and the techniques through which they aim to alleviate and negotiate its consequences. Much like the ways states experience insecurity, so do human beings, even if the latter embody a tangibility, a mortality, and a materiality that exceed nation-states.

In the process of exposing state failures and wrong turns, it became clear that human insecurity and the transgressions on the body resulting from crossing borders, constructions of legality, and illegality and bureaucratic politics were only a few of the effects of a more seriously flawed sovereign machine, the layers of which I was able to explore only in minor detail in this work. The complex set of interactions that determine the production of meaning when it comes to trafficking in human beings, for example, expose not only the layered showings of security as performed by victims, survivors, government officials, social workers, activists, and onlookers alike but also the mutually constitutive instances of once validation/legitimation and then rejection/repression that both victims and state officials embody. In this process of mitigation, the nation-state is not denied or outright rejected, despite the critiques waged at it, but is embodied in the struggles of its most vulnerable populations, depending on the latter for its claim on the legitimate use of violence. What is more, the vulnerable populations discussed in this book, the risks they carry, and the risks that carry them over borders, seas, and mountain ranges tease out the ambiguities inherent in the workings of systems of care and protection without really resolving them. Perhaps this unresolved tension is the stuff that allows for the kind of thinking outside the box that compliments, rather than precludes, engagement with difference.

It is important to note that a clear focus and privileging of insecurity as a permanent state of being has shaped this discussion almost to the detriment of what my students call "the positive stuff of politics." Despite various proclamations relating to the undoing of the state by the sheer existence and persistence of "exceptional life" all around us, a similar engagement with the requisite and inevitable response on the part of sovereign systems has not been offered. The examples engaged with in this book point to processes of constitution of political identities, though not ones always derived from overtly political activities such as protests, marches, and vocal disagreements. Indeed, all cases engaged with in this book engage with populations that exist on the margins and whose silence, rather than outspokenness, define them in their insecurity but also in their agency. The largely invisible survivors of trafficking in human beings, the diseased children of the War on Terror, and the Tibetan refugees living their entire lives in exile do not arrest one's attention because of their visibility or acts of heroic dissent like the example of self-immolation of Tibetan monks in China. Yet, they demand engagement precisely because they can go without being engaged, precisely because their identity is defined by their own struggles to figure out who they are, and because of their either imposed or inherited invisibility, which takes various forms that all relate to and relay a "*politics* of the governed."[1]

As has already been discussed in the chapters so far, the intention of the project was to construct a narrative engaging the struggles of everyday people to navigate borders and in the process, challenge their power, narratives, and barbed wires. Truth be told, I did not encounter a single barbed wire in the process of working on this book, and this fact was unsettling in itself. But I did find iterations of the "barbed wires" all around us, in the figure of Peter the survivor in trafficking, in the stories of the "wise man" in Pokhara, Nepal, and in the lifeless bodies of children on Turkish beaches. This proliferation of the security apparatus and the very real, tangible, material ways in which it intrudes on, cuts through, and mutates the bodies of individuals caught on the outside of territories, for no migrant is ever just in the middle, both unsettles questions of identity and reveals "definitional boundary inscriptions"[2] as flawed. Yet occupying a precarious place of being nowhere, neither in Bulgaria nor in Italy, neither in Nepal nor in Tibet, neither in Turkey nor in Syria, and a status of illegality actually confers upon people the power to interrupt state making and substitute it instead with practices of home, rather than state, making, evidenced in the stories of Bulgarian survivors of trafficking in human beings and Tibetan refugees in Nepal.

On a personal note, I have always been a storyteller and a story keeper. I believe stories can change lives and that individual stories can inform an understanding of what politics, power, and privilege mean better than most other mediums. But stories are also difficult, personal, revealing, and complicated, and they can often leave us feeling broken, confused, used, paranoid, and disappointed. Individual stories, especially ones involving trauma, are even trickier. They are, by definition, also singular and thus unique, unrepeatable, and nongeneralizable, which makes the story in this book one I have never told before. I apologize for the stumbling and fumbling that came about while telling it. It was sometimes unclear to me whether I actually had the right to tell it, and sometimes this right had to be earned repeatedly. This also means I have probably given credit to too few points of reference, am lacking in substantive engagement with the academic mainstream, and that I have held onto just a few points too many on the horizon to find my bearings.

Point one: the beginning. During the research I conducted in Bulgaria in the spring of 2017 as part of my sabbatical leave, I interacted with survivors of trafficking in human beings, government officials, social workers, and representatives from various governmental, nongovernmental, and supragovernmental organizations. I went to Bulgaria, known as one of the top sending countries of trafficked people in Europe, so I could gather material for a book.

Point two: what I expected to find out. I went to Bulgaria with the expectation that I would gain easy access to trafficking victims and that I would be able to gather deeply authentic personal stories that would automatically translate into novel research. As someone whose legitimacy had been secured by a PhD, tenure, and a dozen years of teaching and presenting research at professional meetings, I was confident that I would be able to get the information I needed and do wonders. But also novel wonders.

Armed with my two children, I landed in Bulgaria a day before the biggest snowstorm in the country's winter season hit. No worries, as a Gunnison, Colorado, import, no snowstorm was going to scare me; I had even brought a balaklava. On the third day of being housebound because no snowplows had bothered to show up at our village, my parents' dog bit my son on the head and we had to make our way to the nearest emergency room, which was twenty miles away. The hospital was there, it would provide care, assurance, and a guarantee that my son would be alright, but we could not get to it, not fast at least . . . and by the time we did, D had fallen asleep from the pain and the fear. As my son's head was being stitched without anesthetic (they had run out), in a room where an older man was having a heart attack, and as I sat on top of him trying to keep him from punching the doctor in the head, I started to worry a little. All this did not seem like it was going according to plan and here I was, with no plan B.

Point three: what I found out. Survivors of trauma lie. I soon realized lying was a smart thing to do. Lying got one access to services, money, housing, and maybe a temporary job. Lying was survival, as my Tibetan friends put it, and who could blame them for it? Lying was tricking a system set up to trick you first, getting an upper hand, and being in control. Lying was a type of freedom. Lying often made a story more authentic, as they had lost or sacrificed their truths years ago. I also realized that surviving and coping were about living with lies, broken promises, and that standing to gain from the invention of different stories may, sometimes, give one a head start.

Point four: how to walk away from a "war body." I wake up every morning thinking about Aylan Kurdi, the Syrian toddler who drowned in the Mediterranean. I think of him not only because he is dead and his death was tragic. He is not the "war body" on which to hang our humanitarian/philanthropic angst so as to figure out who we are. But he was alive and, as such, demands attention. Because his death mocks sovereign power by displacing it, too, to the shores of Turkey, it also serves to "reformulate the language of governmentality."[3]

The practices and logics of sovereign power thus had to be abandoned halfway through the project and substituted for stories and experiences of "togetherness." The Tibetan experience, while the most difficult one to describe, taught just that through various practices of home making, mirrored in some way by the desires of survivors of trafficking in human beings in Bulgaria. The bodies of diseased children, while firmly lodged within a state logic of protection point to one of the most effective ways of exceeding said logic: through death. What all three examples share is their temporal removal from dominant sovereign narratives and the ways in which their meaning exceeds both the exception and the law. It is this move that refuses to trivialize their experiences and endows them with deep political meaning, going so far as to destabilize "the entire architecture of how international mobility is governed."[4]

Yet this project is also about missed opportunities, failures of engagement, and barriers of language—my own barbed wires. It does not proclaim to give authoritative pronouncement on the state of all survivors of trafficking in human beings, for example, though it does bring together several case studies which, in their right, interrogate and challenge the art and doing of politics as such. Though I could not resolve the Tibetan case, it helped uncover politics in unusual places and its focus on a politics of waiting pointed to the political potential for resistance contained in the silences, absences, and margins of human lives. The trafficking example, on the other hand, while envisioned as a commentary on the "coping" strategies of survivors, evolved into an examination of governmental policies and academic constructions that, in the case of the former, led to an engagement with the pathologization inherent in migration and the status of a trafficking survivor and, in the case of the latter, to the survivors' continued victimization. The "war bodies" of dead children, in their resolute silence, could not speak for themselves. If this project has managed to do one thing, I hope it has insisted on a "politics of emotion" and visibility as a challenge to oppressive power. I am pretty sure the outrage that Aylan Kurdi's body caused will be outdone by another atrocity; that trafficking in the context of the Eastern European state will continue to feed the financial appetites of entrepreneurial businessmen and sex consumers; and that Tibetans will continue their resistance against cultural assimilation while combating the apathy and precarity of waiting. But what I also hope for, at this juncture, is that this project has initiated a conversation about the possibility of embracing the indeterminate, unruly, messy, and unpredictable nature of human interactions with war, trauma, and risk through idioms such as those of waiting, emotion, and poetry.

NOTES

1. Peter Nyers and Carolina Moulin, "'We Live in a Country of UNHCR'—Refugee Protests and Global Political Society," *International Political Sociology* 1 (2007): 362.
2. Ibid., 362.
3. Ibid., 366.
4. Ibid., 370.

Bibliography

Achter, Paul. "Unruly Bodies: The Rhetorical Domestication of Twenty-First-Century Veterans of War." *Quarterly Journal of Speech* 96, no. 1 (2010): 46–68.

Ackleson, Jackson. "Constructing Security on the US-Mexico Border." *Political Geography* 24, no. 2 (2005): 165–84.

Agamben, Giorgio. *Homo Sacer: Sovereign Power and Bare Life.* Stanford, CA: Stanford University Press, 1998.

———. *Means Without End.* Minneapolis: University of Minnesota Press, 2010.

———. *Remnants of Auschwitz: The Witness and the Archive.* New York: Zone Books, 2002.

———. *State of Exception.* Chicago: University of Chicago Press, 2005.

Alderson, Priscilla. "UN Convention on the Rights of the Child: Some Common Criticisms and Suggested Responses." *Child Abuse Review* 9, no. 6 (2000): 439–43.

Allain, Jean. "No Effective Trafficking Definition Exists: Domestic Implementation of the Palermo Protocol." *Albany Government Law Review* 7 (2014): 111–42.

Allen, Rodney. "Torture, Criminality and the War on Terror." *Alternative Law Journal* 30, no. 5 (2005): 214–18.

Amnesty International. *The Human Cost of Fortress Europe.* London: Amnesty International, 2014.

Amoore, Louise. *The Politics of Possibility.* Durham, NC: Duke University Press, 2013.

Amoore, Louise, and Marieke de Goede. *Risk and the War on Terror.* New York: Routledge, 2008.

Anderson, Bridget. "Stuck in Traffic: How Helpful Is the Trafficking Framework," 2013, www.compass.ox.ac.uk. Accessed October 5, 2017.

Andreas, Peter. *Border Games: Policing the U.S.-Mexico Divide.* Ithaca, NY: Cornell University Press, 2009.

Andrijasevic, Rutvica. "Beautiful Dead Bodies: Gender, Migration and Representation in Anti-Trafficking Campaigns." *Feminist Review* 86 (2007): 24–44.

———. "The Difference Borders Make: (Il)Legality, Migration and Trafficking in Italy among Eastern European Women in Prostitution." In *Uprootings/Regroundings: Questions of Home and Migration*, edited by Sara Ahmed, Claudia Castada, Anne-Marie Fortier, and Mimi Sheller, 251–72. Oxford: Berg Publishers, 2003.

———. *Migration, Agency and Citizenship in Sex Trafficking*. London: Palgrave, 2010.

Appadurai, Arjun. *Fear of Small Numbers: An Essay on the Geography of Anger.* Durham, NC: Duke University Press, 2006.

Aradau, Claudia. *Rethinking Trafficking in Women.* London: Palgrave McMillan, 2008.

Aradau, Claudia, and Rens Van Munster. "Governing Terrorism through Risk." *European Journal of International Relations* 13, no. 1 (2007): 89–115.

Arendt, Hannah. *The Portable Hannah Arendt.* New York: Penguin Classics, 2003.

Baehr, Peter. *Hannah Arendt, Totalitarianism and the Social Sciences.* Stanford, CA: Stanford University Press, 2010.

Balibar, Étienne. *Politics and the Other Scene.* London: Verso, 2002.

Bangali, Shashank, and Sultan Faizy. "Record Number of Children Killed in Afghanistan Conflict in 2016, U.N. Says." *Los Angeles Times*, January 4, 2018, http://www.latimes.com/world/la-fg-afghanistan-report-children-killed-20170206-story.html.

Barnes, Jamal. "The 'War on Terror' and the Battle for the Definition of Torture." *International Affairs* 30, no. 1 (2015): 102–24.

Basaran, Tugba. "Security, Law, Borders: Spaces of Exclusion." *International Political Sociology* 2 (2008): 339–54.

Bauman, Zygmunt. *Globalization: The Human Consequences.* New York: Columbia University Press, 2000.

———. "On Postmodern Uses of Sex." *Theory, Culture and Society* 15 (1998): 19–33.

———. *Wasted Lives: Modernity and Its Outcasts.* London: Polity, 2004.

Bayoumi, Moustafa. "War Stories from Soldiers and Body Washers." *The Progressive* 79, no. 2 (2015): 44–47.

Beck, Ulrich. *What Is Globalization?* Cambridge: Polity Press, 1999.

———. *World Risk Society*. Cambridge: Polity Press, 1998.

———. *World at Risk.* Cambridge: Polity Press, 2009.

Beck, Ulrich, and Elizabeth Beck-Gernsheim. *Individualization: Institutionalized Individualism and Its Social and Political Consequences.* London: Sage, 2001.

Becker, Jo. *Campaigning for Children: Strategies for Advancing Children's Rights.* Stanford, CA: Stanford University Press, 2017.

Bellamy, Alex. "No Pain, No Gain? Torture and Ethics in the War on Terror." *International Affairs* 82, no. 1 (2006): 121–48.

Bernstein, Elizabeth. "Militarized Humanitarianism Meets Carceral Feminism: The Politics of Sex, Rights, and Freedom in Contemporary Antitrafficking Campaigns." *Journal of Women in Culture and Society* 36, no. 1 (2010): 45–71.

———. *Temporarily Yours.* Chicago: University of Chicago Press, 2007.

Biernoff, Suzannah. *Portraits of Violence: War and the Aesthetics of Disfigurement.* Ann Arbor: University of Michigan Press, 2017.

———. "The Rhetoric of Disfigurement in First World War Britain." *Social History of Medicine* 24, no. 3 (2011): 666–85.

Bierstecker, Thomas, and Cynthia Weber. *State Sovereignty as Social Construct.* Cambridge: Cambridge University Press, 1996.

Bigo, Didier. "Frontier Controls in the European Union: Who Is in Control?" In *Controlling Frontiers: Free Movement Into and Within Europe*, edited by Didier Bigo and Elspeth Guild, 49–99. Aldershot: Ashgate, 2005.

———. "Security and Immigration: Toward a Critique of the Governmentality of Unease." *Alternatives* 27 (2002): 63–92.

Bosniak, Linda. *The Citizen and the Alien: Dilemmas of Contemporary Membership.* Princeton, NJ: Princeton University Press, 2008.

Brown, Wendy. *Edgework: Critical Essays on Knowledge and Politics.* Princeton, NJ: Princeton University Press, 2005.

———. "Neo-Liberalism and the End of Liberal Democracy." *Theory and Event* 7, no. 1 (2003): 1–15.

———. *States of Injury.* Princeton, NJ: Princeton University Press, 1995.

———. *Walled States: Waning Sovereignty.* London: Zone Books, 2010.

Bumiller, Kristin. *In an Abusive State.* Durham, NC: Duke University Press, 2008.

Burchell, Graham, Colin Gordon, and Peter Miller, eds. *The Foucault Effect: Studies in Governmentality.* Chicago: University of Chicago Press, 1991.

Butler, Judith. *Frames of War: When Is Life Grievable?* London: Verso, 2016.

———. "Performativity, Precarity and Sexual Politics." *AIBR: Revista de Antropología Iberoamericana* 4, no. 3 (2009): i–xiii.

———. *Precarious Life: The Powers of Mourning and Violence*. London: Verso, 2004.

Butler, Judith, and Athena Athanassiou. *Dispossession: The Performance in the Political.* London: Polity, 2013.
Camayd-Freixas, Erik. "Interpreting after the Largest ICE Raid in US History: A Personal Account." *Latino Studies* 7, no. 1 (2009): 123–39.
Carden-Coyne, Ana. "Masculinity and the Wounds of the First World War: A Centenary Reflection." *French Journal of British Studies* 10, no. 1 (2015): 1–8.
Carpenter, Charli. "'Women and Children First': Gender, Norms and Humanitarian Evacuation in the Balkans 1991–95." *International Organization* 57, no. 4 (2003): 661–94.
Carter, Donald, and Heather Merrill. "Bordering Humanism: Life and Death on the Margins of Europe." *Geopolitics* 12, no. 2 (2007): 248–64.
Caruth, Cathy. *Listening to Trauma.* Baltimore, MD: Johns Hopkins University Press, 2014.
Cavarero, Andriana. *For More Than One Voice: Toward a Philosophy of Vocal Expression.* Stanford, CA: Stanford University Press, 2005.
Chew, Dolores. "Gender, Migration and Trafficking—An Introduction." *Labour, Capital and Society* 39, no. 2 (2006): 1–18.
Cojocaru, Claudia. "My Experience Is Mine to Tell: Challenging the Abolitionist Victimhood Framework." *Anti-Trafficking Review* 7 (2006): 12–36.
Collins, Victoria. *State Crime, Women, and Gender.* London: Routledge, 2016.
Connolly, William. *Why I Am Not a Secularist.* Minneapolis: University of Minnesota Press, 1999.
Curti, Giorgio Hadi. "From a Wall of Bodies to a Body of Walls: Politics of Affect/Politics of Memory/Politics of War." *Emotion, Space and Society* 1 (2008): 106–18.
Das, Veena. *Life and Words: Violence and the Descent into the Ordinary.* Oakland, CA: University of California Press, 2006.
Dauphinée, Elizabeth, and Christina Masters, eds. *The Logics of Biopower and the War on Terror.* London: Palgrave Macmillan, 2007.
Debrix, François, and Alexander Barder. "Nothing to Fear but Fear." *International Political Sociology* 3, no. 4 (2009): 398–413.
Derrida, Jacques. *On Cosmopolitanism and Forgiveness.* New York: Routledge, 2001.
Dillon, Michael. "Governing Terror." *International Political Sociology* 1 (2007): 7–28.
———. *Politics of Security.* London: Routledge, 1996.
———. "Poststructuralism, Complexity and Poetics." *Theory, Culture and Society* 17, no. 5 (2000): 1–26.
———. "The Sovereign and the Stranger." In *Sovereignty and Subjectivity*, edited by Jenny Edkins, Nalini Persram, and Véronique Pin-Fat, 117–40. Boulder, CO: Lynn Reiner, 1999.
Ditmore, Melissa, and Marjan Wijers. "The Negotiations on the UN Protocol on Trafficking in Persons." *Nemesis* 4 (2003): 79–88.
Dobreva, Natasha. *Protection of the Rights of Victims of Trafficking in Human Beings.* Sofia, Bulgaria: Foundation Association Animus, 2013.
Doezema, Jo. "Loose Women or Lost Women? The Re-Emergence of the Myth of White Slavery in Contemporary Discourses on Trafficking in Women." *Gender Issues* (Winter 2000): 23–50.
———. "Ouch! Western Feminists' 'Wounded Attachment' to the 'Third World Prostitute.'" *Feminist Review* 67 (2001): 16–38.
———. *Sex Slaves and Discourse Masters: The Construction of Trafficking.* London: Zed Books, 2010.
Doty, Roxanne Lynn. *Anti-Immigrantism in Western Democracies.* London: Routledge, 2003.
———. "Maladies of Our Souls: Identity and Voice in the Writing of Academic International Relations." *Cambridge Review of International Affairs* 17, no. 2 (2004): 377–92.
———. "States of Exception on the Mexico-US Border: Security, 'Decisions,' and Civilian Border Patrol." *International Political Sociology* 1, no. 2 (2007): 113–37.
Dreufys, Hubert, and Paul Rabinow, eds. *Beyond Structuralism and Hermeneutics.* Chicago: University of Chicago Press, 1983.
Edkins, Jenny. *Face Politics.* London: Routledge, 2015.
———. *Trauma and the Memory of Politics.* Cambridge: Cambridge University Press, 2003.

Ellerman, Antje. "Undocumented Migrants and Resistance to the Liberal State." *Politics and Society* 38, no. 3 (2010): 408–29.

Engbersen, Godfried. "The Unanticipated Consequences of Panopticon Europe: Residence Strategies of Illegal Migrants." In *Controlling a New Migration World*, edited by Virginie Fuiraudon and Christian Joppke, 222–46. London: Routledge, 2001.

Enloe, Cynthia. *Bananas, Beaches and Bases: Making Feminist Sense of International Politics.* Oakland, CA: University of California Press, 2014.

———. *Manuevers: The International Politics of Militarizing Women's Lives.* Oakland, CA: University of California Press, 2000.

Epstein, Charlotte. "Embodying Risk: Using Biometrics to Protect the Borders." In *Risk and the War on Terror*, edited by Louise Amoore and Marieke de Goede, 178–93. New York: Routledge, 2008.

———. "Guilty Bodies, Productive Bodies, Destructive Bodies." *International Political Sociology* 1, no. 2 (2007): 149–64.

Eyben, Rosalind, and Rebecca Napier-Moore. "Choosing Words with Care? Shifting Meanings of Women's Empowerment in International Development." *Third World Quarterly* 30, no. 2 (2009): 285–300.

Felman, Shoshana, and Dori Laub. *Testimony: Crises of Witnessing in Literature, Psychoanalysis and History.* New York: Routledge, 1992.

Ferguson, James. *The Anti-Politics Machine.* Minneapolis: University of Minnesota Press, 1994.

Fitzpatrick, Peter. "Bare Sovereignty." In *Politics, Metaphysics, and Death*, edited by Andrew Norris, 49–73. Durham, NC: Duke University Press, 2005.

Foteinos, Dimitris. "Images of the Body: The Green Physical Education Curriculum Since the Second World War." *History of Education* 41, no. 6 (2012): 807–22.

Foucault, Michel. *Discipline and Punish: The Birth of the Prison.* 2nd edition. New York: Vintage Books, 1995.

———. "Governmentality." In *The Foucault Effect: Studies in Governmentality*, edited by G. Burchell, C. Gordon, and P. Miller, 87–104. Chicago: University of Chicago Press, 1991.

———. *Herculine Barbin.* New York: Vintage, 1980.

———. *The History of Sexuality: An Introduction.* Volume 1. New York: Vintage Books, 1990.

———. *Power/Knowledge: Studies in Governmentality.* Chicago: University of Chicago Press, 1980.

———. *Security, Territory, Population: Lectures at the Collège de France, 1977–1978.* New York: Palgrave, 2007.

———. *Society Must Be Defended: Lectures at the Collège de France, 1975–1976.* New York: Picador, 2003.

———. "The Subject and Power." *Critical Inquiry* 8, no. 4 (1982): 777–95.

Gadda, Andressa. "Rights, Foucault, and Power: A Critical Analysis of the United Nation Convention on the Rights of the Child." *Edinburgh Working Papers in Sociology*, no. 31 (2008): 1–13.

Giddens, Antony. *The Consequences of Modernity.* Stanford, CA: Stanford University Press, 2011.

Gordon, Colin. "Government Rationality: An Introduction." In *The Foucault Effect: Studies in Governmentality*, edited by G. Burchell, C. Gordon, and P. Miller, 1–52. Chicago: University of Chicago Press, 1991.

Halpin, Hayley. "Children Suffered a Shocking Scale of Violence in Conflict Zones in 2017, UNICEF Warns." *The Journal.ie*, January 4, 2018, http://www.thejournal.ie/unicef-children-conflict-2017-3771913-Dec2017/.

Henry, Doug. "Violence and the Body." *Medical Anthropology Quarterly* 20, no. 30 (2006): 379–98.

Hernandez, Julie Gerk. "The Tortured Body, the Photograph, and the U.S. War on Terror." *CLCWeb: Comparative Literature and Culture* 9, no. 1 (2007): 2–10.

Herr, Michael. "Dispatches." In *Unclaimed Experience: Trauma, Narrative and History*, by Cathy Caruth. Baltimore, MD: Johns Hopkins University Press, 1996.

Innes, Alexandria. *Migration, Citizenship and the Challenge for Security.* London: Palgrave Macmillan, 2015.

Jabri, Vivienne. "War, Security and the Liberal State." *Security Dialogue* 37, no. 1 (2006): 47–64.

Jansson, Dominika Borg. *Modern Slavery: A Comparative Study of the Definition of Trafficking in Persons.* Leiden: Brill Nijhoff, 2014.

Jardine, Gail McNicol. *Foucault and Education.* New York: Peter Lang, 2005.

Kahlon, Rajkamal. "Did You Kiss the Dead Body?: Visualizing Absence in the Archive of War." *Comparative Studies of South Asia, Africa and the Middle East* 34, no. 2 (2014): 336–63.

Khanna, Ranjana. "Disposability." *Differences* 20, no. 1 (2009): 181–98.

Kinney, Edith. "Victims, Villains, and Valiant Rescuers: Unpacking Sociolegal Constructions of Human Trafficking and Crimmigration in Popular Culture." In *The Illegal Business of Human Trafficking*, edited by Maria João Guia, 87–108. New York: Springer, 2015.

Kligman, Gail, and Stephanie Limoncelli. "Trafficking Women After Socialism: To, Through and From Eastern Europe." *Social Politics: International Studies in Gender, State and Society* 12, no. 2 (2005): 118–40.

Kottman, Paul. "Introduction." In *For More Than One Voice: Toward a Philosophy of Vocal Expression*, edited by Adrianna Cavarero, vii–xxiv. Stanford, CA: Stanford University Press, 2005.

Kovic, Ron. *Born on the Fourth of July*. Brooklyn: Akashic Books, 2005.

Lennard, Debra. "Censored Flesh." *The British Art Journal* 12, no. 2 (2011): 22–33.

Lindstrom, Nicole. "Regional Sex Trafficking in the Balkans: Transnational Networks in Enlarged Europe." *Problems of Post-Communism* 51, no. 3 (2004): 45–52.

Malkki, Liisa. *Purity and Exile: Violence, Memory, and National Cosmology among Hutu Refugees in Tanzania.* Chicago: University of Chicago Press, 1995.

Mbembé, Achille. "Necropolitics." *Public Culture* 15, no. 1 (2003): 11–40.

McBride, Keally, and Annick T. R. Wibben. "The Gendering of Counterinsurgency in Afghanistan." *Humanity* 3, no. 2 (2012): 199–215.

McConnell, Fiona. *Rehearsing the State.* Oxford: Wiley Blackwell, 2016.

McIvor, David. "Bringing Ourselves to Grief: Judith Butler and the Politics of Mourning." *Political Theory* 40, no. 4 (2012): 409–36.

McLoughlin, Daniel. "Giorgio Agamben on Security, Government and the Crisis of Law." *Griffith Law Review* 21, no. 3 (2012): 680–707.

McNeill, Maggie. "Mid-Witness Testimony: The Unreliability of First Person Accounts in Sex Trafficking Discourse." *Albany Government Law Review* 7 (2014): 56–90.

McSorley, Kevin. "Toward an Embodied Sociology of War." *The Sociological Review* 62, no. 2 (2014): 107–28.

Muller, Benjamin J. *Security, Risk and the Biometric State.* New York: Routledge, 2010.

Munro, Vanessa. "Of Rights and Rhetoric: Discourses of Degradation and Exploitation in the Context of Sex Trafficking." *Journal of Law and Society* 35, no. 2 (2008): 240–64.

National Commission on Combating Trafficking in Human Beings 2016 Report. Sofia, Bulgaria.

National Commission on Combating Trafficking in Human Beings 2017 Report. Sofia, Bulgaria.

Nijboer, Jan, and Judith Vocks. "The Promised Land: A Study of Trafficking in Women from Central and Eastern Europe to the Netherlands." *European Journal on Criminal Policy and Research* 8 (2000): 379–88.

Noah, Trevor. *Born a Crime: Stories from a South African Childhood.* New York: John Murray, 2016.

Norris, Andrew. "Introduction: Giorgio Agamben and the Politics of the Living Dead." In *Politics, Metaphysics, and Death*, edited by Andrew Norris, 1–30. Durham, NC: Duke University Press, 2005.

Nyers, Peter, and Christina Moulin. "'We Live in a Country of UNHCR'—Refugee Protests and Global Political Society." *International Political Sociology* 1 (2007): 356–72.

Ogilvie, Bertrand. *L'Homme Jetable.* Paris: Éditions Amsterdam, 2012.

Oliver, Kelly. *Witnessing: Beyond Recognition.* Minneapolis: University of Minnesota Press, 2001.

Omar Mahmoud, Toman, and Christoph Trebesch. "The Economics of Human Trafficking and Labour Migration: Micro-Evidence from Eastern Europe." *Journal of Comparative Economics* 38 (2010): 173–88.

Orbach, Susie. "People in Distress." In *The Politics of Risk Society*, edited by Jane Franklin, 90–98. Oxford: Polity Press, 1998.

Pandolfi, Mariella. "From Paradox to Paradigm: The Permanent State of Emergency in the Balkans." In *Contemporary States of Emergency: The Politics of Military and Humanitarian Interventions*, edited by Didier Fassin and Mariella Pandolfi, 153–72. New York: Zone Books, 2010.

———. "Laboratory of Intervention: The Humanitarian Governance of the Postcommunist Balkan Territories." In *Postcolonial Disorders*, edited by Mary-Jo DelVeccio Good, Sandra Teresa Hyde, Sarah Pinto, and Byron J. Good, 157–86. Berkeley, CA: University of California Press, 2008.

Parashar, Swati. "What Wars and 'War Bodies' Know about International Relations." *Cambridge Review of International Affairs* 26, no. 4 (2013): 615–30.

Passavant, Paul A. "The Contradictory State of Giorgio Agamben." *Political Theory* 35, no. 2 (2007): 147–74.

Pollock, Griselda. "Liquid Modernity and Cultural Analysis." *Theory, Culture and Society* 24 (2007): 111–16.

Prasad, Uma Shankar. "Study of Nepal's Economic Relations with China." *Journal of Development and Administrative Studies* 23 (2015): 23–32.

Rancière, Jacques. "Who Is the Subject of the Rights of Man?" *South Atlantic Quarterly* 103, no. 2/3 (2004): 297–310.

Raulff, Ulrich. "An Interview with Giorgio Agamben." *German Law Journal* 5, no. 5 (2004): 609–14.

Reiss, Matthias. "Bronzed Bodies behind Barbed Wire: Masculinity and the Treatment of German Prisoners of War in the United States during World War II." *Journal of Military History* 69, no. 2 (2005): 475–504.

Report Concerning the Implementation of the Council of Europe Convention on Action against Trafficking in Human Beings by Bulgaria, GRETA (2015), 32.

Rorty, Richard. *Achieving Our Country: Leftist Thought in Twentieth-Century America.* Cambridge, MA: Harvard University Press, 1999.

Rosière, Stéphane, and Reece Jones. "Teichopolitics: Re-Considering Globalization through the Role of Walls and Fences." *Geopolitics* 17, no. 1 (2012): 217–34.

Ruivo, Joana. "The Fragilities of Human Trafficking Victims." In *The Illegal Business of Human Trafficking*, edited by Maria João Guia, 21–28. New York: Springer, 2015.

Russell, Emily. *Reading Embodied Citizenship: Disability, Narrative and the Body Politic.* Piscataway, NJ: Rutgers University Press, 2011.

Salter, Michael. "Theory of / : The Suture and Critical Border Studies." *Geopolitics* 17 (2012): 734–55.

Sanyal, Debrati. "Poetry in a Time of Terror." *Yale French Studies* 175/176 (2014): 102–18.

Scarry, Elaine. *The Body in Pain: The Making and Unmaking of the World.* Oxford: Oxford University Press, 1985.

Schmitt, Carl. *Political Theology: Four Chapters on the Concept of Sovereignty.* Chicago: University of Chicago Press, 2006.

Shapiro, Michael J. *Politics and Time.* London: Polity, 2016.

Soderlund, Gretchen. "The Rhetoric of Revelation: Sex Trafficking and the Journalistic Exposé." *Humanity* 2, no. 2 (2011): 193–211.

Squire, Vicky. *Post/Humanitarian Border Politics Between Mexico and the US.* London: Palgrave McMillan, 2015.

Stein, Maria. "'We' the Subject: The Power and Failure of (In)Security." *Security Dialogue* 37, no. 2 (2006): 187–205.

Stephens, Sharon. "Children and the Politics of Culture in 'Late Capitalism.'" In *Children and the Politics of Culture*, edited by Sharon Stephens, 3–50. Princeton, NJ: Princeton University Press, 1995.

Sturken, Marita. *Tangled Memories: The Vietnam War, the AIDS Epidemic, and the Politics of Remembering.* Berkeley, CA: University of California Press, 2007.

Sylvester, Christine. *Experiencing War*. London: Routledge, 2011.

———. "Experiencing War: A Challenge for International Relations." *Cambridge Review of International Affairs* 26, no. 4 (2013): 669–74.

———. *Feminist International Relations: An Unfinished Journey.* New York: Cambridge University Press, 2002.

———. *Feminist Theory and International Relations in a Postmodern Era.* Cambridge: Cambridge University Press, 1994.

———. *War as Experience: Contributions from International Relations and Feminist Analysis.* New York: Routledge, 2013.

———. "War Experiences/War Practices/War Theory." *Millennium* 40, no. 3 (2012): 483–503.

Tickner, J. Ann. *Gendering World Politics.* New York: Columbia University Press, 2001.

Tidy, Joanna. "Gender, Dissenting Subjectivity and the Contemporary Military Peace Movement in Body of War." *International Feminist Journal of Politics* 17, no. 3 (2015): 454–72.

Till, Karen. *The New Berlin: Memory, Politics, Place.* Minneapolis: University of Minnesota Press, 2005.

Todres, Jonathan. "Law, Otherness, and Human Trafficking." *Santa Clara Law Review* 49, no. 3 (2009).

UN General Assembly. *Convention on the Rights of the Child*, November 20, 1989, Treaty Series, vol. 1577, p. 3, http://www.refworld.org/docid/3ae6b38f0.html. Accessed February 17, 2016.

———. *Convention Relating to the Status of Refugees*, July 28, 1951, United Nations, Treaty Series, vol. 189, p. 137, http://www.refworld.org/docid/3be01b964.html. Accessed November 10, 2017.

Väyrynen, Tarja. "Keeping the Trauma of War Open in the Male Body: Resisting the Hegemonic Forms of Masculinity and National Identity in Visual Arts." *Journal of Gender Studies* 22, no. 2 (2013): 137–51.

Väyrynen, Tarja, and Eeva Puumala. "Bodies of War, the Past Continuous, and (Ar)thythmic Experiences." *Alternatives* 40, no. 3–4 (2015): 237–50.

Vergès, Françoise. "The Age of Love." *Transformation* 47 (2001): 1–17.

Walker, R. B. J. *Inside/Outside: International Relations as Political Theory.* Cambridge: Cambridge University Press, 1992.

Wall, Thomas Carl. "Au Hasard." In *Politics, Metaphysics, and Death*, edited by Andrew Norris, 31–48. Durham, NC: Duke University Press, 2005.

Weitzer, Ronald. "The Social Construction of Sex-Trafficking: Ideology and Institutionalization of a Moral Crusade." *Politics and Society* 35, no. 3 (2007): 447–75.

Žižek, Slavoj. *Demanding the Impossible.* London: Polity, 2013.

———. *Mapping Ideology.* London: Verso, 1994.

———. *The Sublime Object of Ideology.* London: Verso, 1989.

INTERVIEWS, CHAPTER 3 (PSEUDONYMS ARE USED FOR ANONYMITY)

Bulgarian Agency for Refugees representative. Interview by author, Sofia, Bulgaria, April 4, 2017.

Desislava Ivanova, Secretary, National Commission on Combating Trafficking in Human Beings. Interview by author, Sofia, Bulgaria, February 6, 2017.

Antoaneta Vassileva, Foundation "Association Animus," Sofia, Bulgaria, February 6, 2017.

Ani Torozova, Shelter Director, Foundation "Association Animus." Interview by author, Sofia, Bulgaria, February 6, 2017.

Diliana Markova, Legal Council, Foundation "Association Animus." Interview by author, Sofia, Bulgaria, February 6, 2017.
Diliana Markova, Legal Council, Foundation "Association Animus." Interview by author, Sofia, Bulgaria, March 2, 2017.
Ekaterina Ivanova, Board of Directors, P.U.L.S. Foundation. Interview by author, Pernik, Bulgaria, April 4, 2017.
Julia, P.U.L.S. Foundation. Interview by author, April 4, 2017.
Iana Kacarova, Board of Directors, P.U.L.S. Foundation. Interview by author, Pernik, Bulgaria, March 2, 2017.
Tsvetelina Naidenova, Trafficking Division, International Organization for Migration Bulgaria. Multiple interviews by author between February and June 2017.
Dr. Rosanka Venelinova, Executive Director, Center Nadya. Interview by author, Sofia, Bulgaria, May 4, 2017.
Valya Gerasimova, Secretary, National Commission on Combating Trafficking in Human Beings in Montana, Bulgaria. Interview by author, February 20, 2017.
Cvetan, trafficking shelter director. Interview by author, Bourgas, Bulgaria, May 9, 2017.
Ella, survivor of trafficking in human beings. Interview by author, Bulgaria, April 26, 2017.
Patrice, survivor of trafficking in human beings. Interview by author, Sofia, Bulgaria, May 10, 2017.
Raya, survivor of trafficking in human beings. Interview by author, Sofia, Bulgaria, May 10, 2017.
Elena, survivor of trafficking in human beings. Interview by author, Bulgaria, March 28, 2017.
Irina, survivor of trafficking in human beings. Interview by author, April 28 and June 10, 2017.
Maria, survivor of trafficking in human beings. Interview by author, Sofia, Bulgaria, April 3 and June 18, 2017.
Peter, survivor of trafficking in human beings. Interview by author, Sofia, Bulgaria, May 4, 2017.
Ivana and *Stella*, survivors of trafficking in human beings. Interview by author, Bulgaria, May 9, 2017.

INTERVIEWS, CHAPTER 5

Tashi, Atisha Primary School Principal, Jawalakhel Tibetan Refugee Settlement, Kathmandu, Nepal, May 22, 2017.
Jawalakhel Tibetan Refugee Settlement Camp Manager, May 23, 2017.
Tsering, Secretary, Jawalakhel Tibetan Refugee Settlement, May 23, 2017.
Migmar Dorjee, Secretary, Central Tibetan Administration, Kathmandu, Nepal, May 24, 2017.
Karma Jelseng, Remote Areas Settlement Officer, CTA Offices, Kathmandu, Nepal, May 24, 2017.
Kevin Allen, Representative, United Nations High Commissioner for Refugees, Nepal, May 24, 2017.
Paul Norton, Director, International Organization for Migration, Nepal, May 24, 2017.
Carpet Maker, Jawalakhel Tibetan Settlement, May 25, 2017.
Buddhist Nun, Jawalakhel Tibetan Settlement, May 26, 2017.
Kelsang, Primary School Teacher, Bouddhanath Refugee Camp, May 27, 2017.
Kelsang Mother, Bouddhanath Refugee Camp, May 27, 2017.
Kalsang, Science Teacher, Atisha Primary School, Jawalakhel Tibetan Settlement, Kathmandu, Nepal, May 27, 2017.
Karma Sopa, Tashi Ling Camp Manager, Tashi Ling Tibetan Refugee Settlement, Pokhara, Nepal, May 30, 2017.
Principal, Hermann Geiman Elementary School, Tashi Ling Tibetan Refugee Settlement, Pokhara, Nepal, May 30, 2017.
"Wise Man," Tashi Ling Tibetan Refugee Settlement, Pokhara, Nepal, May 31, 2017.
Sonam, Tashi Ling Tibetan Refugee Settlement, Pokhara, Nepal, May 31, 2017.

Tashi Pelkhiel Camp Manager, Tashi Pelkhiel Tibetan Refugee Settlement, Pokhara, Nepal, May 31, 2017.
Tensing, Science Teacher, Mount Kailash Primary School, Tashi Pelkhiel Tibetan Refugee Settlement, Pokhara, Nepal, May 31, 2017.
Dolma, Shop Owner, Tashi Ling Tibetan Refugee Settlement, Pokhara, Nepal, June 1, 2017.
Lhakpa Choedon, Accountant, Tashi Ling Tibetan Refugee Settlement, Pokhara, Nepal, June 1, 2017.
Phubu Thinley, Headmaster, Hermann Geiman High School, Pokhara, Nepal, June 1, 2017.
Tsechu Dolma, Mountain Resiliency, Jawalakhel Tibetan Refugee Settlement, Kathmandu, Nepal, June 2, 2017.

WEBSITES AND NEWSPAPER ARTICLES

Chapter 3

European Commission. "Together Against Trafficking in Human Beings." https://ec.europa.eu/anti-trafficking/citizens-corner/trafficking-explained_en. Accessed November 5, 2017.
Human Rights First. "Human Trafficking by the Numbers." https://www.humanrightsfirst.org/resource/human-trafficking-numbers. Accessed November 5, 2017.
International Organization for Migration. "Global Trafficking Trends in Focus." https://www.iom.int/news/un-migration-agency-marks-11th-eu-anti-trafficking-day-new-data-analysis-human-trafficking. Accessed on November 5, 2017.
U.S. Department of State. "Trafficking in Persons Report 2017." https://www.state.gov/j/tip/rls/tiprpt/2017/index.htm. Accessed November 5, 2017.
United Nations. "Protocol to Prevent, Suppress and Punish Trafficking in Persons Especially Women and Children, Supplementing the United Nations Convention against Transnational Organized Crime." http://www.ohchr.org/EN/ProfessionalInterest/Pages/ProtocolTraffickingInPersons.aspx.

Chapter 5

Bhutaneserefugees.com. "History of the Bhutanese Refugee Camps." http://bhutanesrefugees.com/history/. Accessed January 27, 2018.
Central Tibetan Administration, Department of Finance. http://tibet.net/department/finance/.
The Diplomat. "A State within a State: Tibetans in Nepal." https://thediplomat.com/2016/07/a-state-within-a-state-tibetans-in-nepal/. Accessed January 27, 2018.
Global Times. "China Should Offer More Generous Aid to Nepal amid Sino-Indian Border Dispute." http://www.globaltimes.cn/content/1061315.shtml. Accessed February 14, 2018.
Government of Nepal, Ministry of Finance. Documents—IECCD Publications. http://www.mof.gov.np/en/document/?c=65&t=&y=. Accessed February 14, 2018.
Hindustan Times. "Nepal Top Recipient of India Aid with Rs 253 Crore." https://www.hindustantimes.com/india-news/nepal-top-recipient-of-india-aid-with-rs-253-crore/story-0HFcfbgSj9zaGVbuE4gASM.html. Accessed August 24, 2018.
India Times. "India's Aid to Nepal Dwindles, China Now among Major Donors." https://economictimes.indiatimes.com/news/politics-and-nation/indias-aid-to-nepal-dwindles-china-now-among-major-donors/articleshow/51984004.cms. Accessed February 14, 2018.
Ministry of Finance, Government of Nepal. http://www.mof.gov.np/en/document/?c=65&t=&y. Accessed February 14, 2018.
New York Times. "Violence in Tibet as Monks Clash with the Police." http://www.nytimes.com/2008/03/15/world/asia/15tibet.html. Accessed February 10, 2018.
Save Tibet. "High-Level Chinese Visit to Nepal Highlights Difficulties for Tibetan Community." https://www.savetibet.org/high-level-chinese-visit-to-nepal-highlights-difficulties-for-tibetan-community/. Accessed February 14, 2018.

Spotlightnepal.com. "World Refugee Day 2017: Nepal Stands with Refugees." https://www.spotlightnepal.com/2017/06/17/world-refugee-day-2017-nepal-stands-refugees/. Accessed January 27, 2018.

Synergy. "The Tale of One Tibetan Refugee." http://utsynergyjournal.org/2017/05/06/the-tale-of-one-tibetan-refugee-in-nepal-memories-insecurity-and-home/. Accessed January 27, 2018.

Tibet Sun. "Exile Tibetan Parliament Passes Budget of 2,434 Million Rupees." https://www.tibetsun.com/news/2017/03/26/exile-tibetan-parliament-passes-budget-of-2434-million-rupees. Accessed February 26, 2018.

Tibetan Review. "Tibetan Refugees in Nepal." http://www.tibetanreview.net/tibetan-refugees-in-nepal/. Accessed February 1, 2018.

Xinhuanet.com. "Nepal to See China's Assistance for Poverty Alleviation in Border Districts." http://www.xinhuanet.com/english/2017-05/30/c_136326330.htm. Accessed February 14, 2018.

Chapter 6

BBC. "Syria Chemical 'Attack': What We Know." http://www.bbc.com/news/world-middle-east-39500947, https://www.nytimes.com/2017/04/04/world/middleeast/syria-gas-attack.html. Accessed January 14, 2018.

Boredpanda.com. "Artists around the World Respond to Tragic Death of 3-Year-Old Syrian Refugee." https://www.boredpanda.com/syrian-boy-drowned-mediterranean-tragedy-artists-respond-aylan-kurdi/. Accessed February 4, 2018.

Cnn.com. "Omran Daqneesh, Young Face of Aleppo Suffering, Seen on Syrian TV." https://www.cnn.com/2017/06/07/middleeast/omran-daqneesh-syrian-tv-interview/index.html. Accessed February 24, 2018.

The Guardian. "Boy in the Ambulance." https://www.theguardian.com/world/2016/aug/18/boy-in-the-ambulance-image-emerges-syrian-child-aleppo-rubble. Accessed January 5, 2018.

———. "Syrian Death Toll over 500 as Bombing in Eastern Ghouta Continues." https://www.theguardian.com/world/2018/feb/24/syria-death-toll-nears-500-as-eastern-ghouta-bombing-continues. Accessed February 24, 2018.

The Journal.ie. "Children Suffered a Shocking Scale of Violence in Conflict Zones in 2017, UNICEF Warns." January 4, 2018, http://www.thejournal.ie/unicef-children-conflict-2017-3771913-Dec2017/.

The Independent. "Refugee Crisis: 600 Children Have Died Crossing the Mediterranean in 2016." http://www.independent.co.uk/news/world/europe/refugee-crisis-600-children-dead-mediterranean-2016-save-the-children-un-a7342131.html. Accessed February 4, 2018.

International Organization for Migration. "Forced Labour, Modern Slavery and Human Trafficking." http://www.ilo.org/global/topics/forced-labour/lang--en/index.htm. Accessed November 5, 2017.

———. "IOM Counts 3,771 Migrant Fatalities in Mediterranean in 2015." https://www.iom.int/news/iom-counts-3771-migrant-fatalities-mediterranean-2015. Accessed January 5, 2018.

———. "Mediterranean Migrant Arrivals Top 363,348 in 2016; Deaths at Sea: 5,079." https://www.iom.int/news/mediterranean-migrant-arrivals-top-363348-2016-deaths-sea5079. Accessed January 5, 2018.

———. "Missing Migrants." https://missingmigrants.iom.int/region/mediterranean/. Accessed February 4, 2018.

Intrepid Report. "Who Is Responsible for 3-Year-Old Aylan Kurdi's Death?" September 10, 2015. http://www.intrepidreport.com/archives/16226. Accessed February 24, 2018.

Los Angeles Times. "Record Number of Children Killed in Afghanistan Conflict in 2016, U.N. Says." http://www.latimes.com/world/la-fg-afghanistan-report-children-killed-20170206-story.html. Accessed January 4, 2018.

The Telegraph. "Fareed Shawky, the 6-Year-Old Yemeni Boy Who Becomes a Symbol of the War, Dies." http://www.telegraph.co.uk/news/worldnews/middleeast/yemen/11944336/Fa

reed-Shawky-the-6-year-old-Yemeni-boy-who-a-became-symbol-of-the-war-dies.html. Accessed February 10, 2018.

———. "New Photos Emerge of Omran Daqneesh." June 5, 2017. http://www.telegraph.co.uk/news/2017/06/05/newphotos-emerge-omran-daqneesh-boy-became-symbol-aleppos-suffering/. Accessed February 24, 2018.

Time. "Aylan Kurdis's Story: Behind the Most Heartbreaking Photo of 2015." http://time.com/4162306/alan-kurdi-syria-drowned-boy-refugee-crisis/. Accessed February 24, 2018.

United Nations High Commissioner for Refugees. "Mediterranean Death Toll Soars, 2016 Is Deadliest Year Yet." http://www.unhcr.org/en-us/news/latest/2016/10/580f3e684/mediterranean-death-toll-soars-2016-deadliest-year.html. Accessed January 5, 2018.

Index

About the Author

Maria Boikova Struble is Associate Professor of Politics and Government at Western Colorado University.

www.ingramcontent.com/pod-product-compliance
Ingram Content Group UK Ltd.
Pitfield, Milton Keynes, MK11 3LW, UK
UKHW041123071026
15587UKWH00009B/142

9 781786 601223